UWB Technology - Circuits and Systems

Edited by Mohamed Kheir

Published in London, United Kingdom

IntechOpen

Supporting open minds since 2005

UWB Technology - Circuits and Systems
http://dx.doi.org/10.5772/intechopen.77459
Edited by Mohamed Kheir

Contributors
Ritesh Saraswat, Ritesh Kumar Saraswat, Patrick Longhi, Ernesto Limiti, Sarthak Singhal, Amit Kumar Singh, Symeon Nikolaou, Abdul Quddious, LiTian Wang, Dalia Elsheakh

First published in London, United Kingdom, 2020 by IntechOpen
IntechOpen is the global imprint of INTECHOPEN LIMITED, registered in England and Wales, registration number: 11086078, 7th floor, 10 Lower Thames Street, London, EC3R 6AF, United Kingdom
Printed in Croatia

British Library Cataloguing-in-Publication Data
A catalogue record for this book is available from the British Library

Additional hard and PDF copies can be obtained from orders@intechopen.com

UWB Technology - Circuits and Systems
Edited by Mohamed Kheir
p. cm.
Print ISBN 978-1-78985-063-5
Online ISBN 978-1-78985-064-2
eBook (PDF) ISBN 978-1-83968-412-8

For EU product safety concerns:
IN TECH d.o.o., Prolaz Marije Krucifikse Kozulić 3, 51000 Rijeka, Croatia, info@intechopen.com or visit our website at intechopen.com.

Meet the editor

Mohamed Kheir was born in Cairo, Egypt, in 1977. He received his MSc degree in Communications Technology from the University of Ulm, Ulm, Germany in 2005, and his PhD degree (with honors) in Information Engineering and Technology from the German University in Cairo, Egypt, in collaboration with Magdeburg University, Germany, in 2011. From 2012 to 2015, he was a Lecturer at the Chair of Microwave Engineering, University of Kiel, Germany, where he was involved in several research projects and teaching duties. From October 2015 to October 2019, he has been a Senior RF-Development Engineer with IMS Connector Systems Group, Germany, where he is responsible for the design and development of high-speed and multi-channel connectors for automotive and 5G applications. Since November 2019, he has been working as a Senior EMC Expert at Keysight Technologies, Germany.

His research interests include microwave/mm wave integrated circuits, EMC measurements, and UWB Technology.

Dr. Kheir serves as an Associate Editor at IEEE Access and IEEE Internet of Things Journal. He is also a reviewer for multiple IEEE/IET journals and a TPC-member for many conferences. Dr. Kheir is the recipient of several prestigious awards and honors from different conferences and organizations worldwide.

Contents

Preface

Ultra Wide Band (UWB) technology has attracted increasing interest and there is a growing demand for UWB for several applications and scenarios. The unlicensed use of the UWB spectrum has been regulated by the Federal Communications Commission (FCC) since the early 2000s. The main concern in designing UWB circuits is to consider the assigned bandwidth and the low power permitted for transmission. This makes UWB circuit design a challenging mission in today's community. The major aim of the book is to introduce the readers to the basic design principles and building blocks that are necessary for such technology. In addition, the latest research innovations, especially in the area of circuits and systems, are presented. Various circuit designs and system implementations are described within this book to give the reader a glimpse of the state-of-the-art examples in this field. The book starts at the circuit level design of major UWB elements such as filters, antennas, and amplifiers; and ends with the complete system implementation using such modules.

The first chapter of the book is titled "Frequency Reconfigurable UWB Antenna Design for Wireless Applications" written by R. K. Sarraswat and M. Kumar. The chapter introduces a reconfigurable antenna suitable for WLAN, WiMAX, WiFi, and UWB wireless applications. The antenna is capable of switching between the narrow band, dual band, and UWB modes by using five p-i-n diodes placed inside an inverted L-shaped ground slot. The proposed design provides for easy integration with cognitive radio and multi radio wireless terminal devices. Radiation characteristics of the proposed antenna are achieved with good impedance matching at these resonant frequencies. The radiation pattern, gain, and efficiency are consistent over all the operating bands, making the proposed antenna a good choice for wireless applications.

The second chapter is titled "UWB Circuits and Sub-Systems for Aerospace, Defence, and Security Applications" and is written by E. Limiti and P. E. Longhi. This chapter reviews the topology, technology, and trends of microwave circuits in UWB systems for aerospace and defence applications. This includes front-end and back-end circuits, amplifiers and mixers, ADCs, and chip integration techniques. An outline on multi-functional circuits (Single-Chip Front-Ends and Core-Chips) is also presented.

The third chapter is titled "Inner Tapered Tree Shaped Ultra Wide Band Fractal Antenna with Polarization Diversity" and is authored by S. Singhal. The chapter introduces a compact third iteration fractal antenna for UWB using an orthogonal arrangement of two antenna elements. The designed antenna has wider bandwidth and excellent polarization diversity performance, which makes it suitable for UWB polarization applications in future wireless communication systems to mitigate the multipath fading.

The fourth chapter is titled "Antennas for UWB Applications" and is authored by S. Nikolaou and A. Quddious. The chapter presents selected antenna designs in relation to the UWB applications and their dictating radiation and operation principles. The demonstrated UWB antennas include antennas for handheld devices used for personal area network (PAN) communications, antennas used for localization and positioning, UWB antennas for RFIDs, radar antennas for through-wall imaging, for ground penetrating radar, for breast tumor detection, and more generally, UWB antennas used for sensing. For some of the aforementioned applications, UWB antennas with special characteristics are needed and these are presented in this chapter, with the relevant applications. These include reconfigurable UWB antennas, metamaterial-loaded UWB antennas, and conformal UWB antennas. The usefulness of these special characteristics in comparison with the claimed advantages is critically evaluated.

The fifth chapter is titled "The UWB Bandpass Filters", which is authored by L.-T- Wand, Y. Xiong, and M. He. The chapter reviews the development process and regulations of UWB bandpass filters. Subsequently, it focuses on the application scenarios of UWB filters in UWB communication systems and the unique merits of UWB filters. In addition, the primary performance specifications of the UWB filters, including insertion loss, return loss, the level of out-of-band attenuation, and roll-off-rate are also presented. After a brief discussion of microwave network theory, several methods for implementing UWB filters are summarized. Furthermore, the design of the UWB filter with notch band is presented in the last section of the chapter.

Finally, the last chapter titled "Passive Components for Ultra-Wide Band (UWB) Applications" is authored by D. N. Elsheakh and E. A. Abdallah. This chapter introduces many types of UWB antennas. The first type is the monopole antenna, which is used as linearly polarized antennas. It has proven to be the best choice for use in various automobiles and mobile equipment due to its simplicity in design. Multilayer antenna with multiple feed points are different types used in UWB antennas with moderate gain. Log parodic and Yagi antennas are other types of UWB with high gain. Electromagnetic band gap (EBG) structures as defected ground or split ring resonators are also used to improve the antenna bandwidth and achieve UWB.

Different UWB filters are presented and described in this chapter as well. These filters may be switchable or tunable in order to introduce notch frequencies within the passband to provide interference immunity from unwanted radio signals, such as wireless local area networks (WLAN) and worldwide interoperability for microwave access (WiMAX) that cohabit within the UWB spectrum. With the current scenario of emerging services of the wireless communication systems, the need for compact multiband filtenna (filter combined with antenna) with the ability to cover the current standards at the microwave band and the next generation standards at the millimeter wave band simultaneously is rising. Also, the operation of cognitive radio and self-adaptive systems need to dynamically monitor the frequency spectrum in search of the unused licensed channels. All these applications need different types of filtennas, which may be fixed, switchable, or tunable that can operate in the UWB range of frequency and also with narrow band modes to cover for example UWB/WiMAX applications.

Dr. Mohamed Kheir
Keysight Technologies,
Germany

Chapter 1

Frequency Reconfigurable UWB Antenna Design for Wireless Applications

Ritesh Kumar Saraswat and Mithilesh Kumar

Abstract

A frequency band reconfigurable antenna is designed and experimentally analyzed. Proposed design achieve UWB (3.1–10.6 GHz), two dual band and two narrow band switching states by implementation of PIN diodes inside the slotted ground. Antenna covers the various wireless standards WLAN, WiMAX, WiFi and UWB with return loss $S_{11} < -10$ dB. Proposed antenna is also operating at 9.2 GHz to include the airborne radar applications. Simulated and experimental results are compared and found to be in good agreement.

Keywords: cognitive radio (CR), frequency-band reconfigurable, RF switch p-i-n diode, ultra wideband (UWB) antenna, WLAN

1. Introduction

Present scenario of wireless communication system required compact and multiple band antenna design. Since many systems are operating at multiple frequency range, requiring dual and triple band antenna for various applications such as WLAN, WiMAX, RFID, satellite communication, etc. Presently, many printed monopole antenna are proposed. Serve for wireless applications to cover the wireless standards for Wireless local area network (WLAN: 2.4–2.48, 5.15–5.35, and 5.75–5.825 GHz) and worldwide interoperability for microwave access (WiMAX: 3.4–3.69 GHz) are two among the available wireless standards which allow interconnections of devices for communication. To achieve multi-functionality, various parameters of antenna such as polarization characteristics, resonant frequency, patterns and impedance bandwidth etc., are reconfigured as per requirements [1].

Presently wireless communication systems are adopting the concept of cognitive radio system where using a sensing antenna performing the monitoring of the spectrum, and can be reconfigured to operate over a desired frequency band. This system is required a frequency band reconfigurable antenna as a sensing element [2]. Most of frequency band reconfigurable antennas providing the band switching between narrowband modes [3–5].

Antenna obtained the quad-band switching by implementation of microelectromechanical systems (MEMS) switch [3]. Another frequency band reconfigurable patch antenna is proposed that operate in four different modes with the help of switching elements [4]. Recently, many microstip patch antenna have been designed that indicate the switching facility in narrowband as well as wideband modes [6–11]. In [6], a Vivaldi antenna is designed that provide the wideband and narrowband mode switching facility. A frequency band reconfigurable antenna with four photoconductive switches is proposed that operating with switching between the three narrowband modes and UWB mode [7]. In [9], antenna has been proposed with narrowband and wideband functionality with reconfigurability characteristics is achieved with the implementation of p-i-n and varactor diodes. Tunable EBG structure are analyzed with active switching devices FET and obtained the transmission characteristics of the structure [12, 13]. Many techniques such as defective ground [14], etching slots [15, 16], metamaterial loading [17–23], dielectric resonator [24], fractal geometry [25, 26], etc., are applied to accomplish multiband reconfigurable operation to cover various wireless applications.

In this chapter, firstly design the octagonal shape patch antenna and implementing the inverted L shaped switchable slotted ground yielded switchable resonant modes such as, two narrowband modes (5.05–5.89 and 8.76–9.80 GHz), two dual band modes (2.21–2.52 GHz and 5.07–5.89 GHz and 2.18–2.52 GHz and 8.78–9.71 GHz) and UWB mode (2.87–16.56 GHz) for wireless applications. As per requirement to design antenna to frequency band reconfigurability introducing the five switching elements p-i-n diodes placed inside the slotted ground. The proposed design is compact in size as compared to antennas are discussed in published literature [6–11]. The simulation work of antenna is done by using CST Microwave Studio (CST MWS) software [27] and measurement is performed with the help of VNA (vector network analyzer-E5071C (300 KHz–20 GHz) ENA series Agilent Technologies). The fabrication of proposed structure is executed by using of PCB prototype machine (Caddo-71).

Following sections focused on the antenna designing with parametric study and switchable modes analysis with results in simulated as well as measurement modes.

2. Ultra wideband (UWB) antenna design

In this section initially octagonal shape monopole antenna size of 40 × 40 × 0.40 mm^3 is designed, as represented in configuration "a" of **Figure 1**. The proposed design is constructed on 0.40 mm thick Roger RT 5880 substrate with the relative permittivity 2.2, fed with 50 ohm microstrip feed line. Octagonal shape radiating element has the dimensions ab = ef = 10 mm, bc = ha = 5.14 mm, cd = gh = 6.0 mm and de = fg = 5.14 mm attached with feedline of dimensions 22 × 1.25 × 0.01 mm^3. On back side of antenna, ground plane

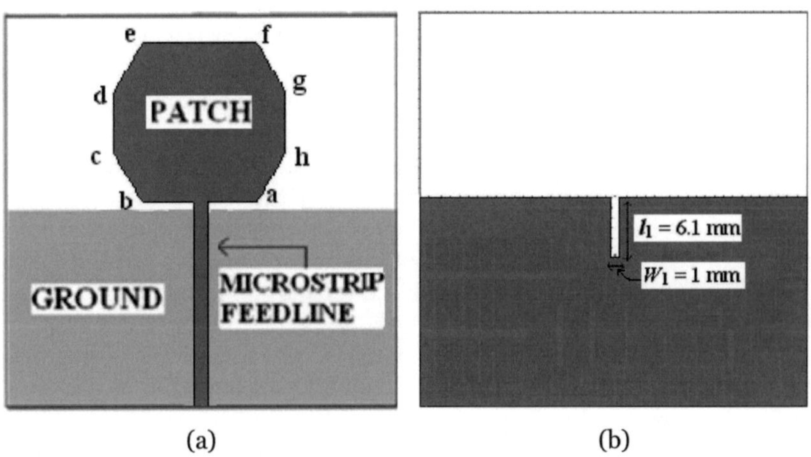

Figure 1.
Configuration of the UWB antenna: (a) front view of structure and (b) back view of structure.

exist with length of 21.1 mm, width of 40 mm and thickness of 0.01 mm. **Figure 1** shows the configuration of the proposed antenna with a top view and bottom view.

The operational performance of the antenna is analyzed with variations in its parameters such as ground slot dimensions (length l_1 and width W_1) are known as the parametric study of the proposed design. This study is carried out by variations in slot length l_1 and width W_1 while keeping other parameters constant. Slot length l_1 and width W_1 is varied from 0 to 7 mm and 0.6 to 1.4 mm respectively, as depicted in **Figure 2**. It is observed that, as variation are done in the value of l_1 and W_1, the reflection coefficient (S_{11}) is changes and the respective frequency band is also changes accordingly. The optimized impedance matching for UWB band characteristics is achieved at length l_1 = 6.1 mm and width W_1 = 1 mm.

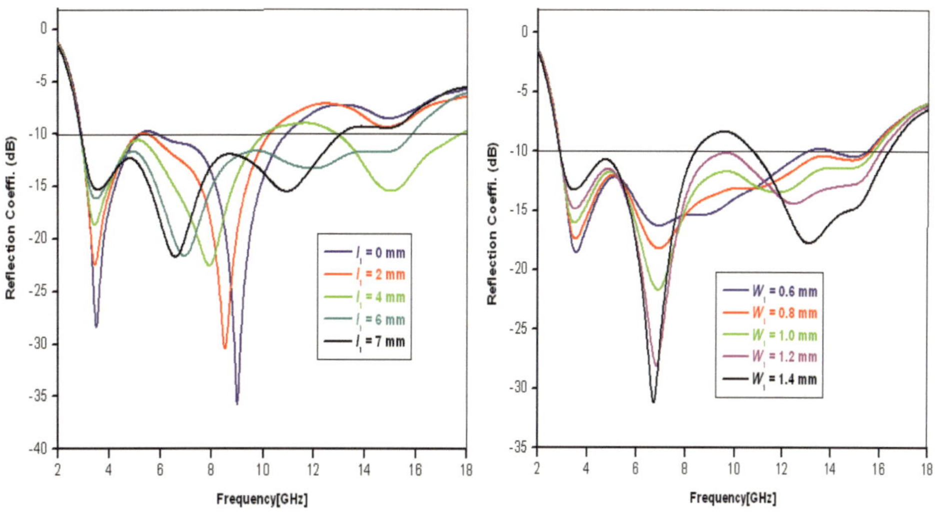

Figure 2.
Simulated reflection coefficient S_{11} of the proposed antenna for different values of l_1 and W_1.

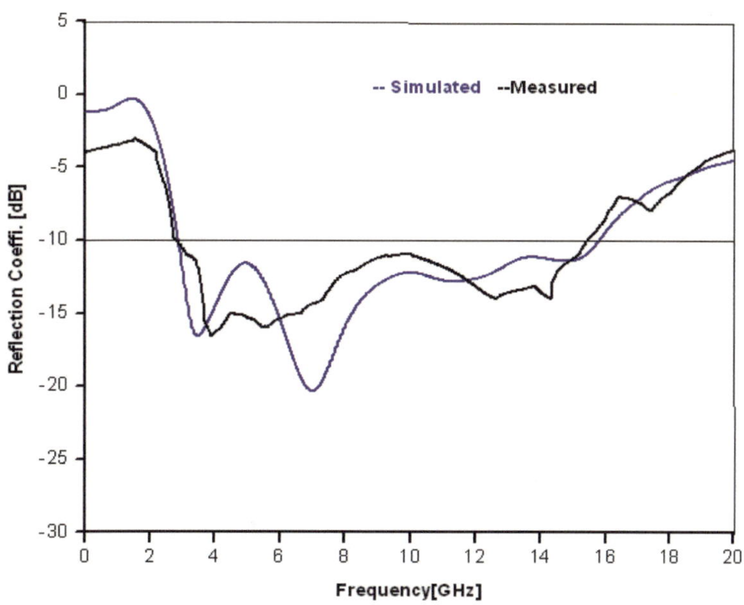

Figure 3.
Simulated and measured reflection coefficient S_{11} of the proposed UWB antenna.

From **Figure 2**, it is indicated that at lower frequencies (2–4 GHz) that the impedance matching is improved when the slot dimensions are reduced (either by reducing l_1 or W_1). At higher frequencies (above 5 GHz), the impedance matching is enhanced when the slot dimensions are increased. The input reflection coefficient S_{11} (below −10 dB) of UWB antenna is achieved at the optimized value of l_1 = 6.1 mm and W_1 = 1 mm. The impedance bandwidth of 141% (2.87–16.56 GHz) under simulation and 140% (2.85–15.85 GHz) in measurement is obtained as shown in **Figure 3**.

3. Narrowband and dual band antenna design

The UWB monopole antenna can be reconfigured to others frequency bands by using an inverted L and rectangular shaped slotted structure placed on the ground plane, as shown in **Figure 4**. This inverted L-shaped slot in ground plane generating an additional current path due to the perturbation of the current flow in antenna structure that leads to the filter characteristics, responsible to suppress the frequencies outside the desired frequency band. These ground slots are generating the stop bands in the UWB frequency range [9]. **Figure 5** represents the different filter structures of the proposed design by variation in the inverted L-shaped slot.

Figure 6 represents the input reflection coefficient S_{11} (below −10 dB) of the antenna for different filter structures. Structure I and II are creating the dual bands whereas structure III and IV are responsible for obtaining the single bands only.

Moreover, the bandwidth of each filtering structure in **Figure 6**, is controllable with changing the length l_2 and width W_2 of parallel vertical arms (in **Figure 4**). By the variations of these arms dimensions impedance bandwidth changes accordingly, as shown in **Figure 7**. While increasing the slot length l_2 and width W_2,

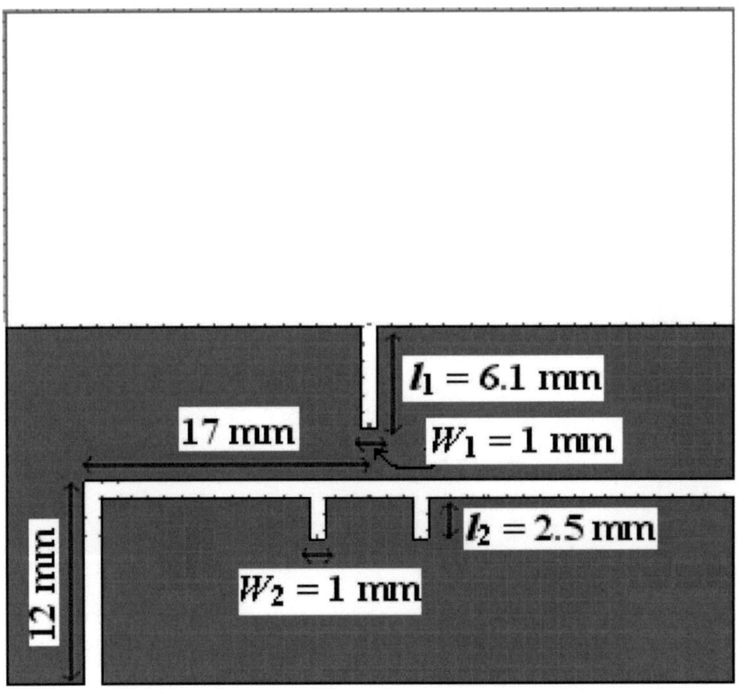

Figure 4.
Slotted structure on ground plane.

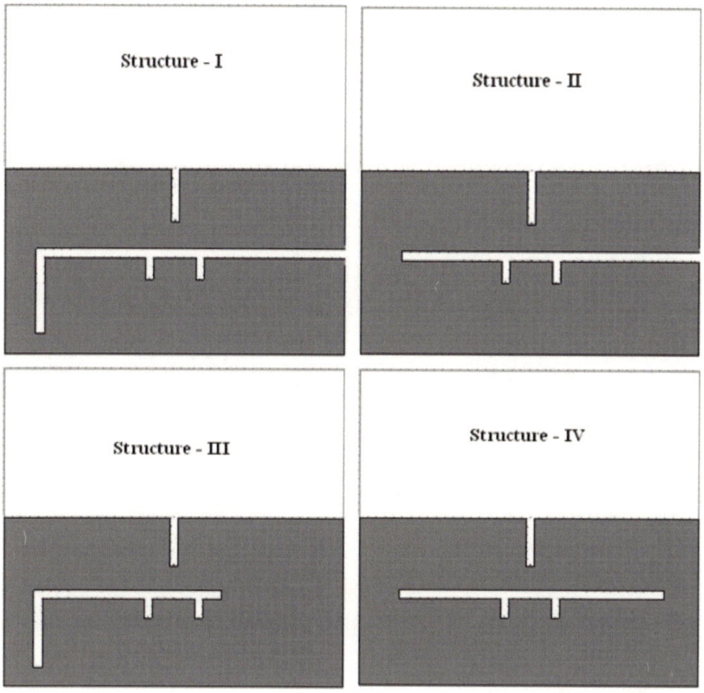

Figure 5.
Filter structures placed on the ground plane.

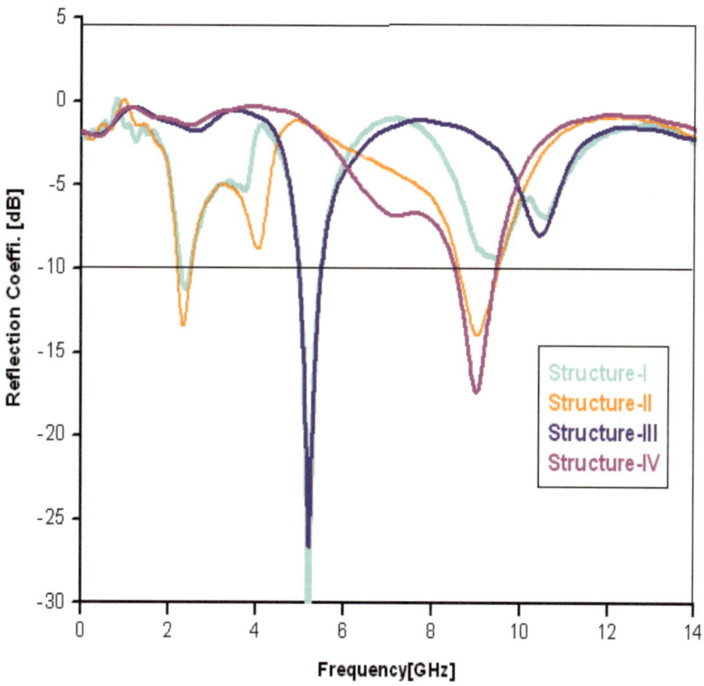

Figure 6.
Simulated reflection coefficient S_{11} of the antenna for filter structures in **Figure 5**.

the bandwidth of antenna decreases from 30 to 12% in structure III. The desired resonant band with the input reflection coefficient S_{11} (below −10 dB) is achieved at the optimized value l_2 = 2.5 mm and W_2 = 1 mm respectively.

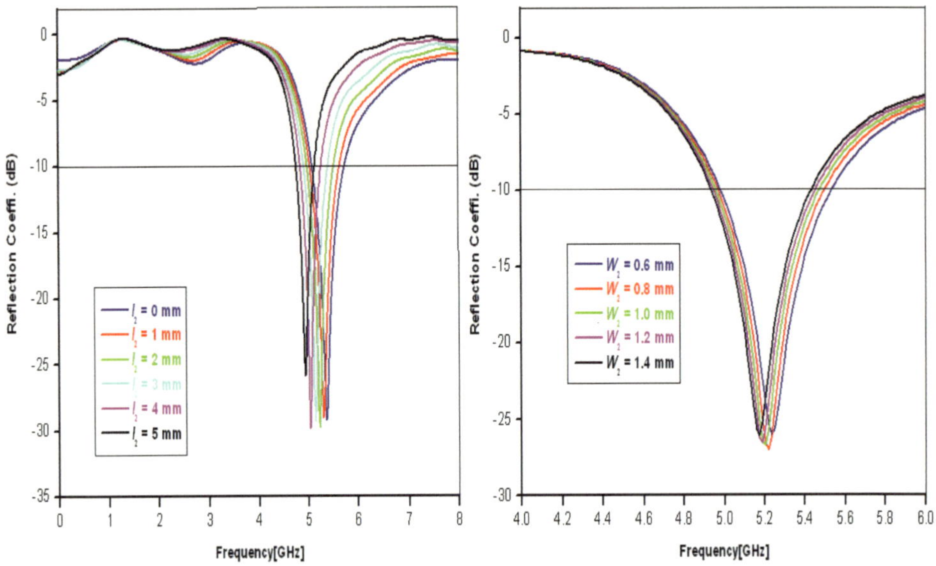

Figure 7.
Simulated reflection coefficient S_{11} of the antenna for different values of l_2 and W_2 in structure-III.

4. Frequency reconfigurable antenna design

In this section analyzed the reconfiguration of UWB mode of proposed design to another narrowband and dual band modes. This reconfiguration is done by implementation of filter structures on the ground plane by placing of five switching elements p-i-n diodes inside it, as indicated in **Figure 8**. The switches D_1–D_5 are positioned in such a way to obtain the required structures I–IV for desired frequency bands. If diode D_1 is on and remaining are off, we will get a filter structure like structure-I for dual band.

For biasing of p-i-n diodes, apply the dc voltage across the p-i-n diodes with the help of metal strips dimension of 2 × 0.6 mm^2, as indicated in **Figure 8**. As shown in **Figure 8**, blocking capacitor of 100 pF is also connected with diode, to provide the isolation between the dc and the RF signal. A beam lead p-i-n diodes (ALPHA-6355) are placed inside the ground slot, where 0.7 V dc is required for biasing of diode. During ON state (forward bias) of diode, it exhibit resistance of 2.6 ohm while in case of OFF state (reverse bias) it represents 0.081 pF [28].

Figure 9 shows the electrical equivalent circuit of the diode for both states (ON/OFF state). For On state, it represents a series combination of fixed inductor (L_s) and a current-controlled resistor (R_s), whereas for OFF state, it indicates a shunt combination of intrinsic-layer capacitance (C_p) and the resistance (R_s) in series with fixed inductance (L_s). The intrinsic-layer capacitance (C_p) is a combination of the stray capacitance C_s and the junction's capacitance C_j.

As per **Table 1**, narrow bands, dual bands and UWB band are obtained by changing the states of diodes and compare the frequency bands and 10-dB bandwidth in simulation and measurement mode. The proposed antenna is initially simulated with the help of simulation software CST Microwave Studio (MWS) [27] and thereafter, fabricated on FR4 substrate with optimized values. **Figure 10** shows the fabricated prototype of the proposed antenna.

A setup is used for frequency band reconfigurable structure to observe the measured reflection coefficient (S_{11}) with the help of vector network analyzer (VNA) and radiation characteristics by using anechoic chamber, shown in **Figure 11**.

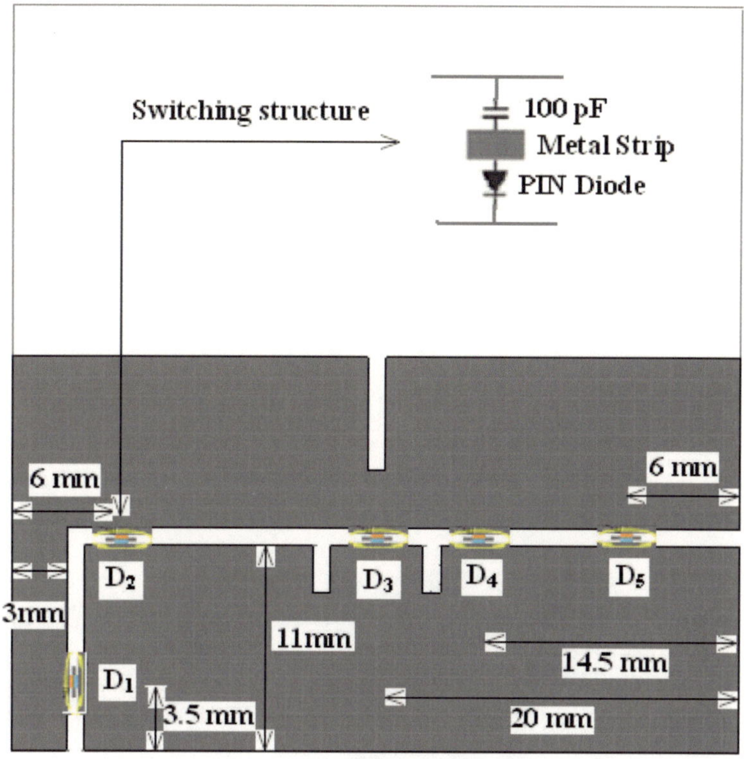

Figure 8.
Switchable filter structure on the ground plane (unit: millimeters).

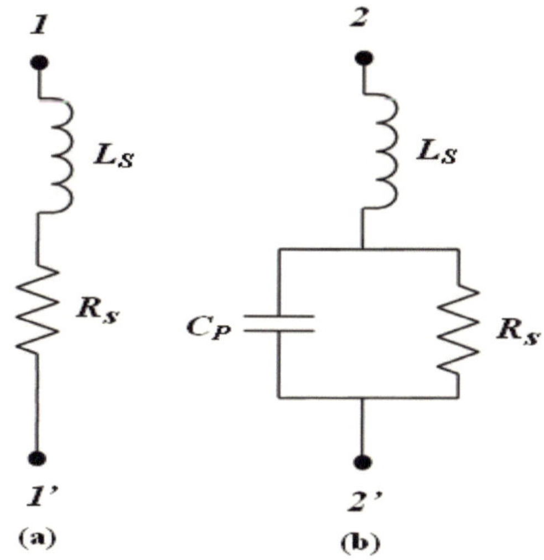

Figure 9.
Equivalent circuit for p-i-n diode: (a) ON-state (forward bias) and (b) OFF-state (reverse bias).

The simulated and measured reflection coefficients S_{11} for all five states are shown in **Figure 12**. Comparison of simulated S_{11} with measured ones is indicated as a good agreement between them. From **Table 1**, for narrowband states I achieve the bandwidth of 16% (5.05–5.89 GHz) and 14% (5.01–5.79 GHz) in simulation and measurement mode respectively. For state II (narrowband), obtained bandwidth of 11% (8.76–9.80 GHz) and 10% (8.68–9.69 GHz) in simulation and measurement

Diode	D_1	D_2	D_3	D_4	D_5	Frequency bands (in GHz)		10-dB bandwidth (%)		Characteristics
States						Simulated	Measured	Simulated	Measured	
I	ON	OFF	OFF	ON	OFF	5.05–5.91	5.01–5.79	16	14	Narrow band
II	OFF	ON	OFF	OFF	ON	8.76–9.80	8.68–9.69	11	10	Narrow band
III	ON	OFF	OFF	OFF	OFF	2.21–2.52 and 5.07–5.89	2.20–2.50 and 5.05–5.90	13 and 15	12 and 15	Dual band
IV	OFF	ON	OFF	OFF	OFF	2.18–2.52 and 8.78–9.71	2.19–2.50 and 8.70–9.60	14 and 10	13 and 9	Dual band
V	ON	ON	ON	ON	ON	2.87–16.56	2.85–15.85	141	140	UWB

Table 1.
Details of combinations of p-i-n diodes with simulated and measured frequency band and bandwidth in each states.

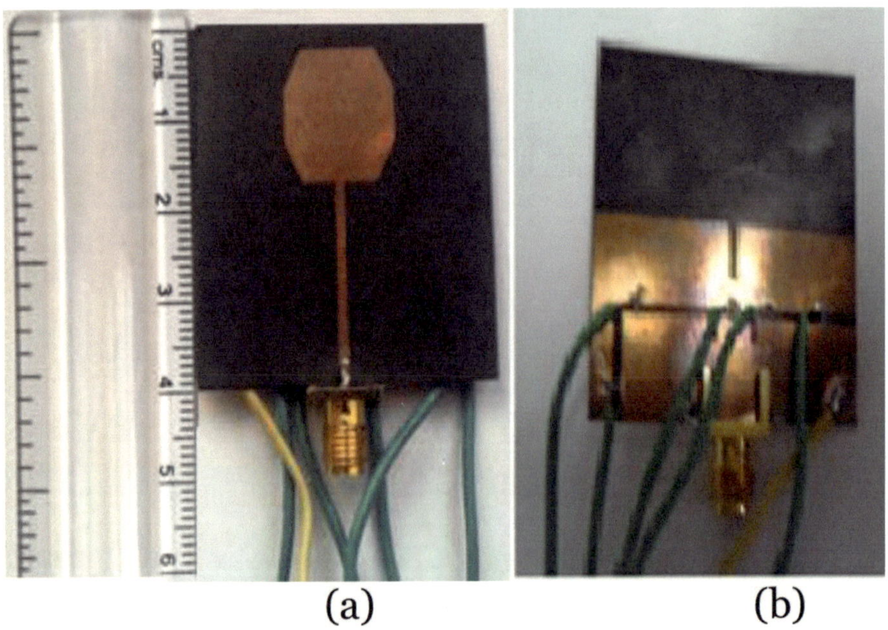

Figure 10.
Images of the fabricated antenna: (a) top view and (b) bottom view.

mode respectively. For next state III, antenna resonant in dual band mode and achieve impedance bandwidth of 13% (2.21–2.52 GHz) and 15% (5.07–5.89 GHz) under simulation and 12% (2.20–2.50 GHz) and 15% (5.05–5.90 GHz) during measurement. For State IV, antenna identifies the operating bandwidth of 14% (2.18–2.52 GHz) and 10% (8.78–9.71 GHz) and 13% (2.19–2.50 GHz) and 9% (8.70–9.60 GHz) during simulation and measurement mode respectively. For V state of UWB mode, antenna indicates the operating bandwidth of 141% (2.87–16.87 GHz) and 140% (2.97–16.80 GHz) under simulation and measurement mode respectively.

The resonant bands are achieved by switching states of diodes can serve several wireless applications such as WLAN, WiMAX, WiFi and UWB. As per IEEE standards the WLAN is identify for 802.11b/g/n (2.4–2.48 GHz), 802.11a/h/j/n (5.2 GHz) and ISM band (2.4–2.5 GHz). Wireless standards WiMAX, WiFi and UWB are identify for frequency bands of 2.3–2.4 and 5.15–5.85 GHz, 2.40–2.48

DOI: http://dx.doi.org/10.5772/intechopen.86035

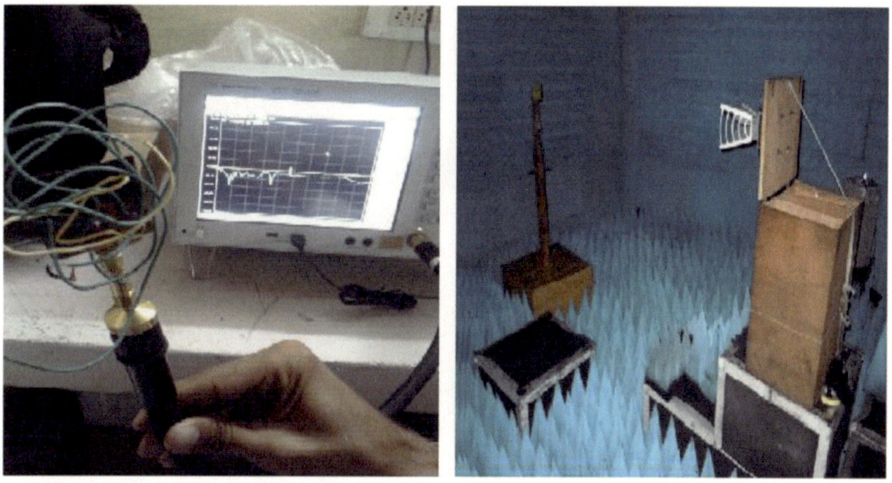

Figure 11.
Images of measurement setup for proposed antenna.

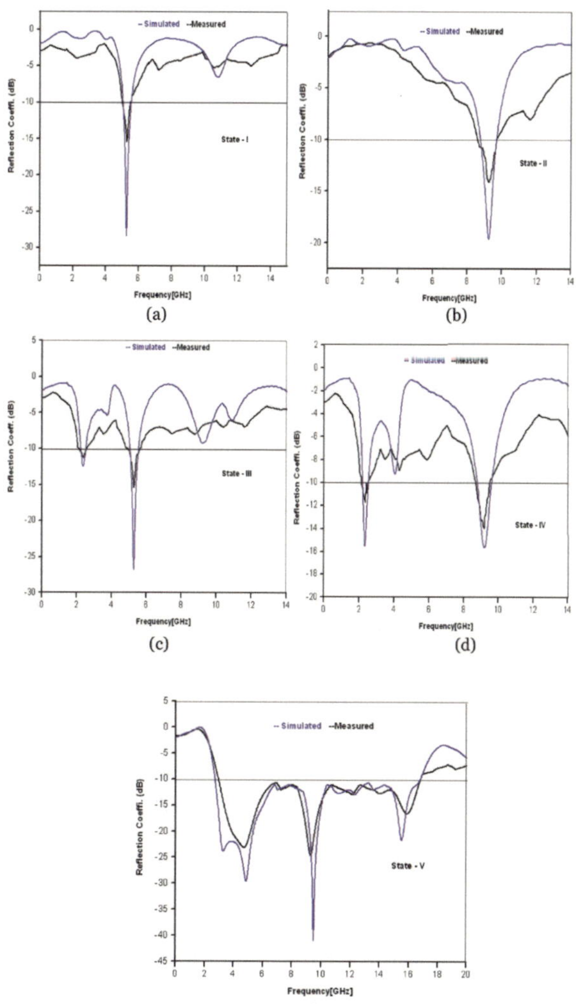

Figure 12.
Simulated and measured reflection coefficient S_{11} of the proposed antenna for states I-V (from (a)-(e) as per **Table 1***).*

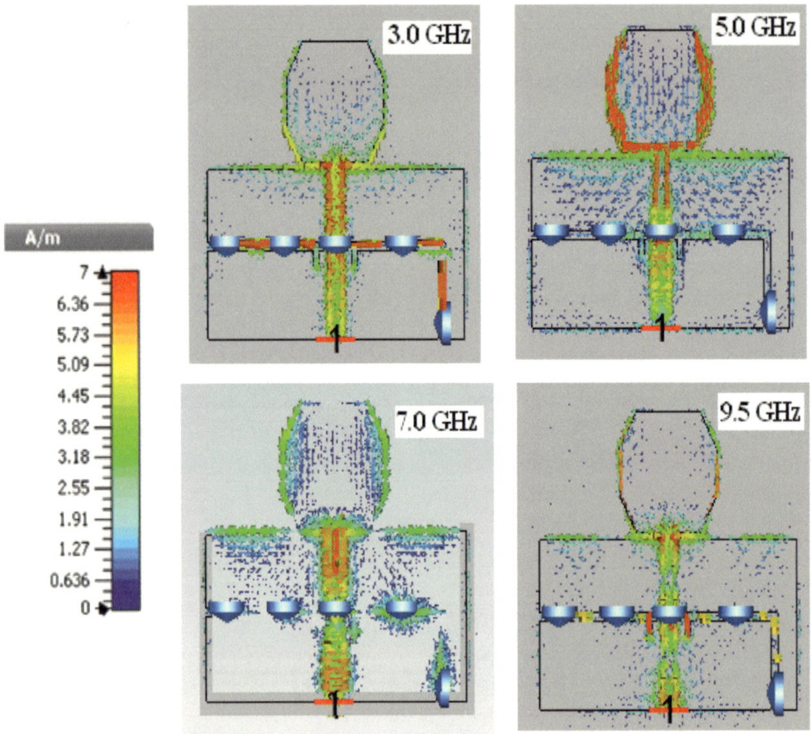

Figure 13.
Surface current distribution of the proposed antenna for different frequencies.

and 5.15–5.85 GHz and 3.1–10.6 GHz respectively. Proposed design also covers the airborne radar applications works at 9.2 GHz.

From **Figure 12(e)**, the resonance is identified at the frequency of 3.0, 5.0, 7.0 and 9.6 GHz. As per the observation of **Figure 13**, it is found that the first resonance is controlled by the inverted L shaped slot dimensions since the maximum surface current is present across it. Second resonance 5.0 GHz is obtained due to the octagonal shape of radiating element and the feedline attached to the patch. Third resonance at 7.0 GHz is obtained due to the rectangular slot created on the ground plane. The two rectangular slits dimensions of l_2 and W_2 are responsible to generate tank circuit causes the fourth resonance at 9.5 GHz. The surface current distribution is observed at theses resonance frequencies as shown in **Figure 13**.

From **Figure 14**, the 3D-gain of the antenna is observed at different resonant frequencies, where the maximum radiation is identify at the various values of angles (theta and phi). It is noticed that at higher frequency, the directivity is improved so that the gain is increased. **Figure 15** represents the measured antenna gain in single band, dual band and UWB modes for various switching states. It is analyzed that at lower frequency range gain is reduced whereas at higher frequencies (above 6 GHz) gain is improved as compared to reference gain level of 4 dB. It is also observed that antenna exhibit the acceptable gain in narrowband and dual-band modes. **Figure 15(b)** shows the average gain of 3.9 dB is achieved for UWB mode of the proposed antenna.

Figure 16(a) indicates the variation of the simulated radiation efficiency from 96.9 to 79.5% and from 97.2 to 70.3% for switching states I and II respectively, for the proposed antenna. The variation of the simulated radiation efficiency from 95.5 to 73.3% and from 94.0 to 78.1% is observed for state III and IV respectively. **Figure 16(b)** shows the simulated radiation efficiency variation from 98.9 to 85.8% for UWB

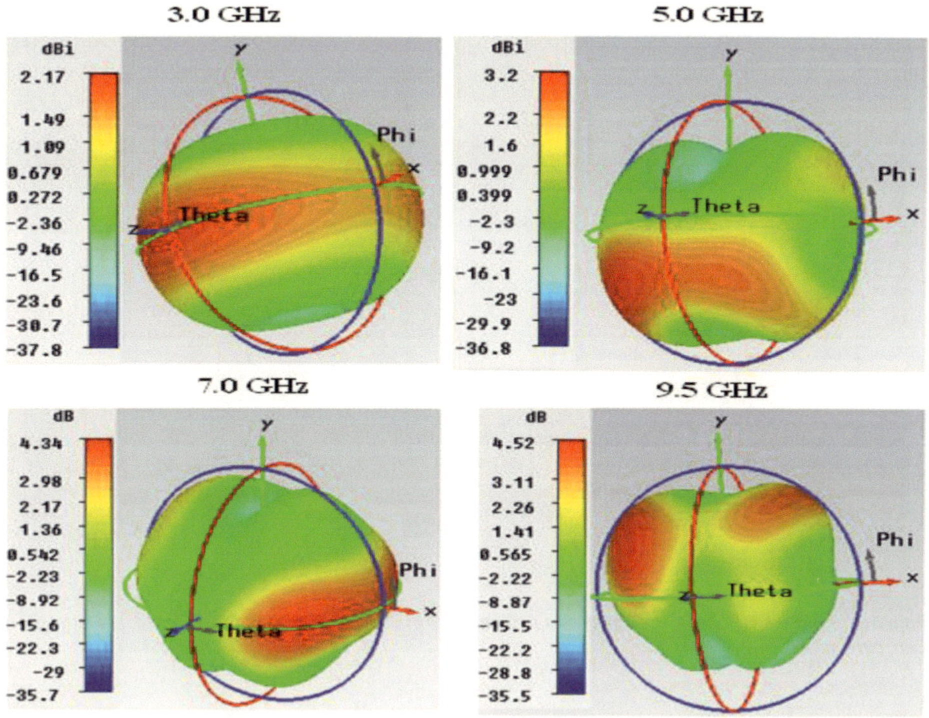

Figure 14.
Simulated gain (dB) of the proposed antenna for different frequencies.

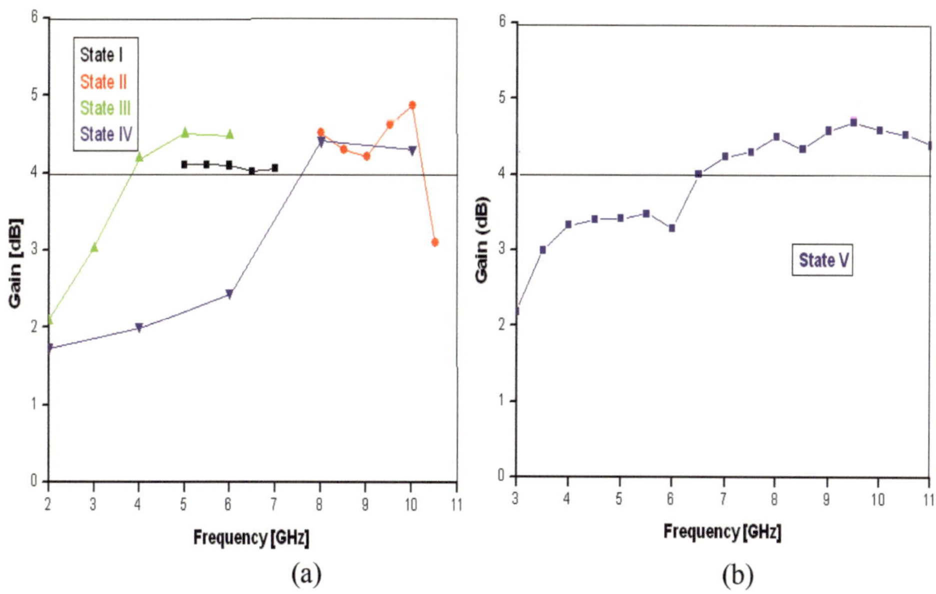

Figure 15.
Measured gain of the proposed antenna for different switching states: (a) single-band and dual-band modes and (b) UWB.

mode. It is noticed that the radiation efficiency is stay above the 70% in all the narrow band, dual band and UWB band. Another observation is that at higher frequency range the simulated radiation efficiency is decreases.

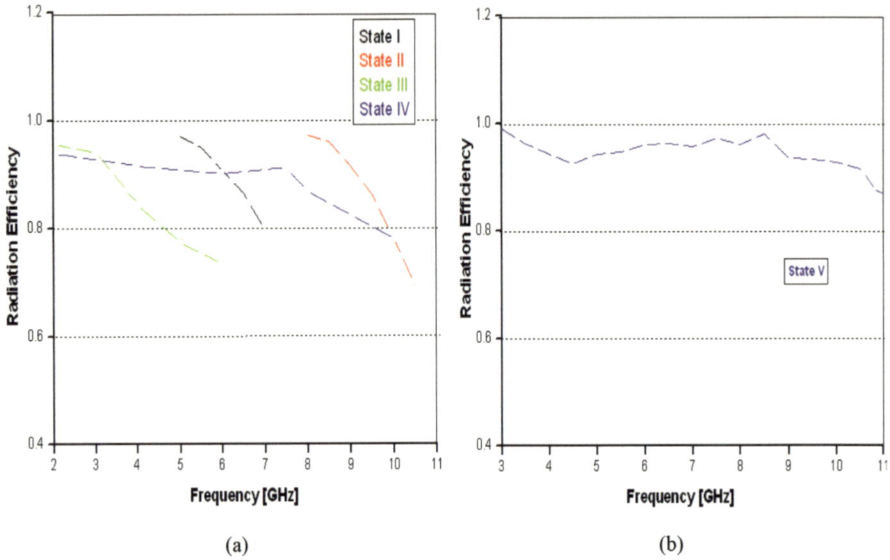

Figure 16.
Simulated radiation efficiency of the proposed antenna for different switching states: (a) single-band and dual-band modes and (b) UWB.

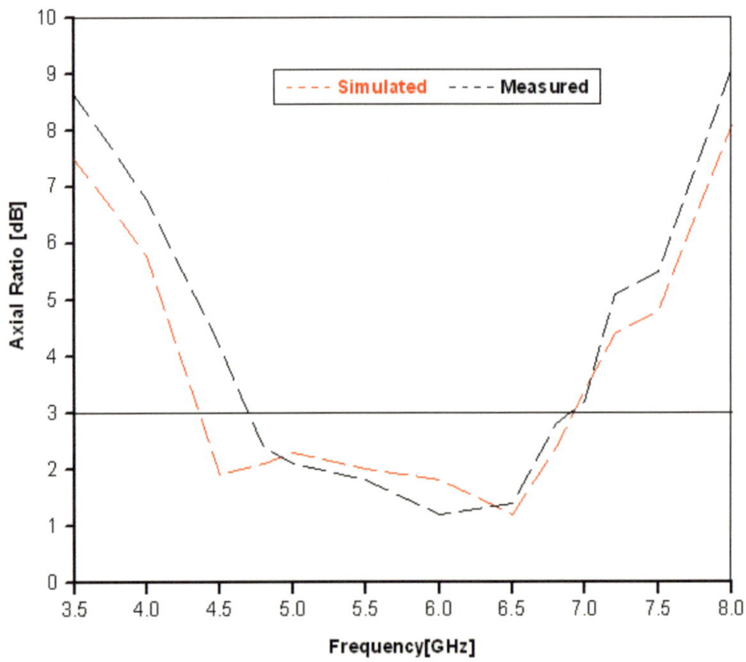

Figure 17.
Simulated and measured AR (axial ratio) (along $\theta = 78°$ and $\Phi = -89°$) of the proposed antenna.

The axial ratio (AR) measurement of the proposed antenna is done inside an anechoic chamber by using antenna measurement system with VNA. To obtain maximum ARBW (axial ratio bandwidth), the antenna measurement system is aligned along the directions of $\theta = 78°$ and $\Phi = -89°$, where AR is stay below the 3 dB reference level. The simulated ARBW of 38% is achieved for frequency range from 4.65 to 6.85 GHz as shown in **Figure 17**. The measured ARBW is slightly less than the simulated one at the center frequency 5.65 GHz.

Patterns are analyzed at operating frequencies 2.4, 5.4, 7.5, and 10 GHz for E and H plane (principal plane). From **Figure 18**, there is dumb bell shape and quasi-omnidirectional like radiation patterns in H-Plane and E-Plane respectively, which represents that the proposed design is a good candidate for wireless communication. There is a good agreement seen between the measured and simulated radiation patterns for E and H plane with the slight difference caused due to assembly

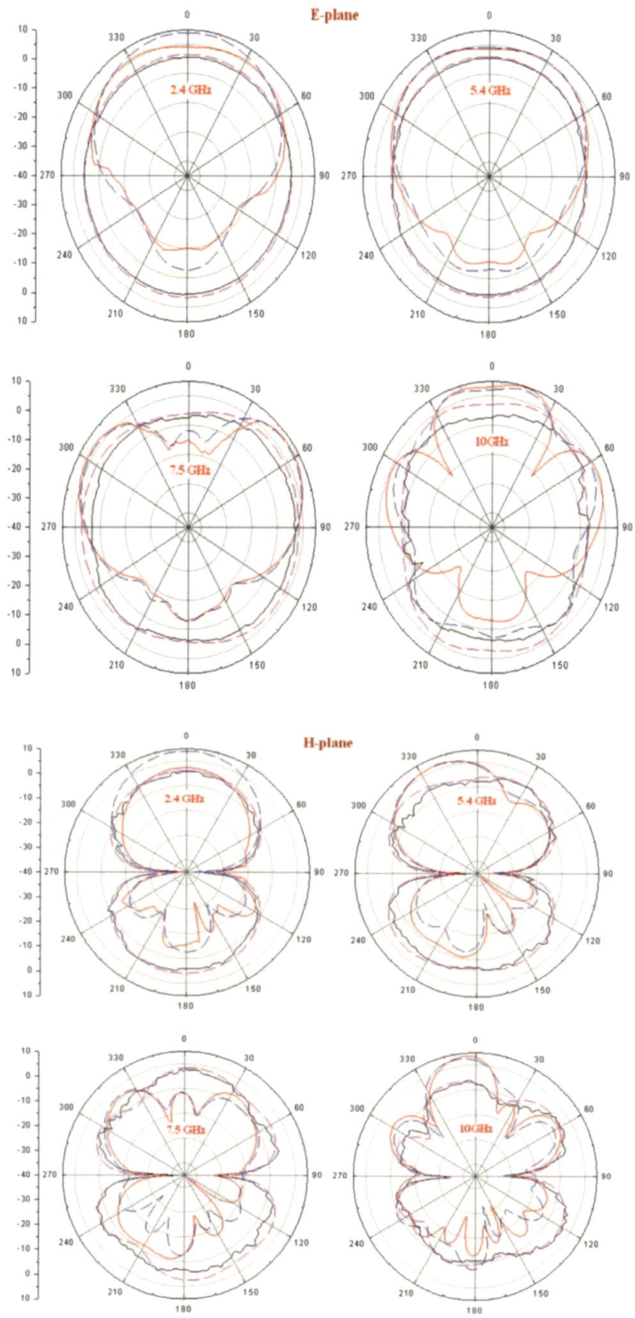

Figure 18.
Measured and simulated E and H plane radiation patterns.

Ref.	Year	Antenna size (in mm^3)	Bandwidth (in %)	Gain (in dBi)	Radiation efficiency (in %)	No. of operating modes (in GHz)	Covered wireless standards	Frequency Band Reconfigurability Achieved
Ref [17]	2013	49 × 49 × 6	2.63/3.28/6.44	4.93/2.85/5.12	89.9/91.8/97.6	3	GPS/WLAN/WiMAX (1.52/2.44/3.57)	No
Ref [20]	2013	52.6 × 30 × 1	47.27/38.88	−0.56/−0.62	89.2/98.1	2	GPS/WLAN (1.5/2.4)	No
Ref [21]	2014	40 × 40 × 1.6	34.48/18.28/19.96	3.97/4.04/3.25	—	3	(WLAN/WiMAX (2.4/3.5/5.8)	No
Ref [22]	2015	56 × 44 × 0.8	5.56/5.86/19.34/13.69	1.3/2.3/3.5/4.4	76.8/80.1/96.6/85.5	4	GPS/WLAN/WiMAX (1.5/2.4/3.5/5.4)	No
Ref [14]	2015	55 × 50 × 1.9	25.3/16.95/12.32	5.71/6.16/6.48	79/86.6/88.7	3	(WLAN/WiMAX (2.4/3.5/5.8)	No
Ref [23]	2016	48 × 48 × 1.6	20.73/15.02/31.96	1.64/2.07/4.06	66.2/77.15/87.6	3	GPS/WLAN (1.9/2.4/5)	No
Ref [15]	2017	40 × 40 × 1.6	1.9/14/5	−4.5/3.75/5.3	10/91/87	3	(GPS/WLAN/WiMAX (1.5/3.5/ 5.4)	No
Ref [24]	2017	50 × 50 × 1.6	2/12/18.2	6.35/5.57/3.9	82.1/84.9/90.9	3	(WLAN/ X Band) (4.8/5.8/9.2)	No
Ref [25]	2018	40 × 40 × 1.6	11.49/3.37/8.61	1.78/3.5/4.4	75.62/72/73	3	(WLAN/ WiMAX Band) (2.4/3.5/ 5.5)	No
Ref [26]	2018	30 × 24.8 × 1.6	3.5/5.01/13.2/5.77	1.35/1/1.07/1.75	—	4	(WiMAX/X Band) (3.1/5.52/7.31/9.72)	No
Proposed antenna		40 × 40 × 0.40	13/15/16/141/16/11	2.1/3.3/3.5/3.9/2.2/4.2	96.7/92.7/94.2/74.4/95.8/91.2	6	WLAN/WiMAX/WiFi/UWB/ ISM Band/Radar Application Band (2.4/5.2/5.8/3.1–10.6/2.45/9.2)	Yes

Table 2.
Comparison of propose designed with those in the state-of-art literature.

misalignments. A consistent omnidirectional radiation is observed in the E plane and a nearly bi-directional pattern is observed along the H plane for all the operating frequencies.

Both the measured and simulated E and H plane radiation patterns are appear reasonably stable with respect to resonant frequency. It is also observed that there is pinch-off along the end fire directions (θ = ±90°) at lower frequencies for H-plane pattern. The E-plane pattern shows the unidirectional nature at higher frequencies because at these frequencies the back lobes (along θ = 180°) are considerably decreases. **Table 2** shows the comparison of the proposed antenna characteristics, like as antenna size, impedance bandwidth, gain, radiation efficiency and operating modes, with reported multiband antennas for wireless standard.

5. Conclusion

A frequency band reconfigurable antenna suitable for WLAN (2.4/5.2 GHz). ISM band (2.4–2.5 GHz), WiMAX (2.3–2.4 and 5.15–5.85 GHz), WiFi (2.40–2.48 and 5.15–5.85 GHz) and UWB (3.1–10.6 GHz) wireless standards are presented in this chapter. Proposed design also covers the airborne radar applications works at 9.2 GHz. The radiating element of octagonal shape and switchable slotted ground is implemented to achieve the frequency band reconfigurability between wireless standards. The switching between the narrowband, dual band and UWB modes is obtain by using five p-i-n diodes placed inside the inverted L shaped ground slot. The proposed design is provides the facility of easily integration with cognitive radio and multi radio wireless terminal devices. Proposed design achieve the bandwidth of 16% (5.05–5.89 GHz) and 14% (5.01–5.79 GHz) in simulation and measurement mode respectively for narrowband states I. Next it obtained bandwidth of 11% (8.76–9.80 GHz) and 10% (8.68–9.69 GHz) in simulation and measurement mode respectively for narrowband states II. Antenna resonant in dual band mode and achieve impedance bandwidth of 13% (2.21–2.52 GHz) and 15% (5.07–5.89 GHz) under simulation and 12% (2.20–2.50 GHz) and 15% (5.05–5.90 GHz) during measurement for next state III. For next state IV, antenna identifies the operating bandwidth of 14% (2.18–2.52 GHz) and 10% (8.78–9.71 GHz) and 13% (2.19–2.50 GHz) and 9% (8.70–9.60 GHz) during simulation and measurement mode respectively. For UWB mode of V state, antenna indicate the operating bandwidth of 141% (2.87–16.87 GHz) and 140% (2.97–16.80 GHz) under simulation and measurement mode respectively. The average gain of 3.9 dB is achieved for UWB mode of the proposed antenna. The radiation efficiency is stay above the 70% in all the narrow band, dual band and UWB band. Radiation characteristics of the proposed antenna are achieved with good impedance matching at these resonant frequencies. The radiation pattern, gain and efficiency are consistent over all the operating bands making the proposed antenna a good choice for wireless applications.

Conflict of interest

The author(s) declare(s) that there is no conflict of interest regarding the publication of this paper.

Author details

Ritesh Kumar Saraswat[1*] and Mithilesh Kumar[2]

1 M.L.V. Govt. Textile and Engineering College, Bhilwara, Rajasthan, India

2 Rajasthan Technical University, Kota, Rajasthan, India

*Address all correspondence to: ritesh.saraswat9@gmail.com

References

[1] Zhang C, Yang S, Pan HK, Fathy AE, Nair VK. Frequency reconfigurable antennas for multi radio wireless platforms. IEEE Microwave Magazine. 2009;**10**(1):66-83. DOI: 10.1109/MMM.2008.930677

[2] FCC (Federal Communications Commission). First Report and order. 2002

[3] Li RL, Wu T, Eom SY, Myoung SS, Lim K, Laskar J, et al. Switchable quad-band antennas for cognitive radio base station applications. IEEE Transactions on Antennas and Propagation. 2010;**58**(5):1468-1476. DOI: 10.1109/TAP.2010.2044472

[4] Mahmoud SF, Sheta AF. A widely tunable compact patch antenna. IEEE Antennas and Wireless Propagation Letters. 2008;**7**:40-42. DOI: 10.1109/LAWP.2008.915796

[5] Huang CT, Han TY. Reconfigurable monopolar patch antenna. Electronics Letters. 2010;**46**(3):199-200. DOI: 10.1049/el.2010.3242

[6] Gardner P, Hamid MR, Hall PS, Ghanem F. Switched-band Vivaldi antenna. IEEE Transactions on Antennas and Propagation. 2011;**59**(5):1472-1480. DOI: 10.1109/TAP.2011.2122293

[7] Li RL, Jin GP, Zhang DL. Optically controlled reconfigurable antenna for cognitive radio applications. Electronics Letters. 2011;**47**(17):948-950. DOI: 10.1049/el.2011.1958

[8] Gardner P, Hamid MR, Hall PS, Ghanem F. Vivaldi antenna with integrated switchable band pass resonator. IEEE Transactions on Antennas and Propagation. 2011;**59**(11):4008-4015. DOI: 10.1109/TAP.2011.2164197

[9] Ghafouri-Shiraz H, Tariq A. Frequency-reconfigurable monopole antennas. IEEE Transactions on Antennas and Propagation. 2012;**60**(1):44-50. DOI: 10.1109/TAP.2011.2167929

[10] Gardner P, Kelly JR, Hall PS. Integrated wide-narrow band antenna for switched operation. In: Processing IEEE EuCAP; 2009; Berlin, Germany. 2009. pp. 3757-3760

[11] Boudaghi H, Azarmanesh M, Mehranpour M. A frequency-reconfigurable monopole antenna using switchable slotted ground structure. IEEE Antennas and Wireless Propagation Letters. 2012;**11**:655-658. DOI: 10.1109/LAWP.2012.2204030

[12] Thalakotuna D, Matekovits L, Heimlich M, Esselle KP, Hay SG. Active switching devices in a tunable EBG structure: Placement strategies and modeling. Journal of Electromagnetic Waves and Applications. 2011;**25**(11):1740-1751. DOI: 10.1163/156939311797164873

[13] Thalakotuna DNP, Esselle KP, Matekovits L, Heimlich M, Hay SG. Changing the electromagnetic bandgap and stopbands in a multistate periodic circuit. Microwave and Optical Technology Letters. 2013;**55**(8):1871-1874. DOI: 10.1002/mop.27675

[14] Ahsan MR, Islam M, Ullah MH. Computational and experimental analysis of high gain antenna for WLAN/WiMAX applications. Journal of Computational Electronics. 2015;**14**(2):634-641. DOI: 10.1007/s10825-015-0699-8

[15] Rajabloo H, Kooshki VA, Oraizi H. Compact microstrip fractal Koch slot antenna with ELC coupling load for triple band application. AEU-International Journal of Electronics and Communications. 2017;**73**:144-149. DOI: 10.1016/j.aeue.2016.12.027

[16] Ali T, Fatima N, Biradara RC. Miniaturized multiband reconfigurable fractal slot antenna for GPS/GNSS/Bluetooth/WiMAX/X-band applications. AEU-International Journal of Electronics and Communications. 2018;**94**:234-243. DOI: 10.1016/j.aeue.2018.07.017

[17] Xu HX, Wang GM, Qi MQ. A miniaturized triple-band metamaterial antenna with radiation pattern selectivity and polarization diversity. Progress in Electromagnetics Research. 2013;**137**:275-292. DOI: 10.2528/PIER12081008

[18] Saraswat RK, Kumar M. Miniaturized slotted ground UWB antenna loaded with metamaterial for WLAN and WiMAX applications. Progress in Electromagnetics Research B. 2016;**65**:65-80. DOI: 10.2528/PIERB15112703

[19] Rahimi M, Zarrabi FB, Ahmadian R, Mansouri Z, Keshtkar A. Miniaturization of antenna for wireless application with difference metamaterial structures. Progress in Electromagnetics Research. 2014;**145**:19-29. DOI: 10.2528/PIER13120902

[20] Xu HX, Wang GM, Lv YY, Qi MQ, Gao X, Ge S. Multi frequency monopole antennas by loading metamaterial transmission lines with dual-shunt branch circuit. Progress in Electromagnetics Research. 2013;**137**:703-725. DOI: 10.2528/PIER12122409

[21] Samsuzzaman M, Islam T, Abd Rahman NH, MRI F, Mandeep JS. Compact modified swastika shape patch antenna for WLAN/WiMAX applications. International Journal of Antennas and Propagation. 2014;**2014**: 1-8. DOI: 10.1155/2014/825697

[22] Cao YF, Cheung SW, Yuk TI. A multiband slot antenna for GPS/WiMAX/WLAN systems. IEEE Transactions on Antennas and Propagation. 2015;**63**(3):952-958. DOI: 10.1109/TAP.2015.2389219

[23] Alam T, Samsuzzaman M, Faruque MRI, Islam MT. A metamaterial unit cell inspired antenna for mobile wireless applications. Microwave and Optical Technology Letters. 2016;**58**(2):263-267. DOI: 10.1002/mop.29543

[24] Vinodha E, Raghavan S. Double stub microstrip fed two element rectangular dielectric resonator antenna for multiband operation. AEU-International Journal of Electronics and Communications. 2017;**78**:46-53. DOI: 10.1016/j.aeue.2017.05.020

[25] Varamini G, Keshtkar A, Naser-Moghadasi M. Compact and miniaturized microstrip antenna based on fractal and metamaterial loads with reconfigurable qualification. AEU-International Journal of Electronics and Communications. 2018;**83**:213-221. DOI: 10.1016/j.aeue.2017.08.057

[26] Ali T, Saadh M, Biradar RC. A fractal quad-band antenna loaded with L-shaped slot and metamaterial for wireless applications. International Journal of Microwave and Wireless Technologies. 2018;**10**(7):826-834. DOI: 10.1017/S1759078718000272

[27] Computer Simulation Technology–CST (Microwave Studio MWS). Version-2014

[28] Alpha Industries. ALPHA-6355 beamlead PIN diode. Data sheet [Online]. 2015. Available from: http://www.datasheetarchive.com/ALPHA/PINdiode6355-datasheet.html

Chapter 2

UWB Circuits and Sub-Systems for Aerospace, Defence and Security Applications

Ernesto Limiti and Patrick E. Longhi

Abstract

In order to maintain technological superiority over other systems, modern equipment for aerospace, defence and security (ADS) applications require advanced integrated circuits operating at microwave and millimetre wave frequencies. High integration is necessary to obtain low SWaP-C features thus enabling the installation of this category of equipment in unfriendly environments: compact spaces, and subject to heavy mechanical loads and temperature stress. This chapter reviews the topology, technology and trends of microwave circuits in UWB systems for ADS applications. Amplification at high frequency is a crucial function: high power amplifiers in the transmit (Tx) chain and low-noise amplifiers in the receive (Rx) chain will be revised, in addition to medium-power (gain) amps. Signal conditioning and routing is also essential: MIMO architecture are becoming the standard and therefore switching and signal phasing and attenuation is increasingly needed, to obtain the desired beam steering and shaping. Each type of circuits leverages the benefits of either gallium nitride (GaN) or gallium arsenide (GaAs), and the role of the semiconductor will be explained. Finally, an outline on multi-functional circuits (single-chip front-ends and core-chips) will be presented: the trend is to realize the whole microwave section of a Tx/Rx module with only to MMICs that perform all the functionalities requested at microwave frequencies.

Keywords: microwave front-ends, microwave measurement circuits, microwave transmit/receive modules, multi-functional MMICs, AESA, III–V semiconductors

1. Introduction

High-frequency circuits are necessary to process the microwave signal between the radiating element and the digital signal processing (DSP) unit. Although, analogue-to-digital converters (ADC) are constantly increasing their frequency and power handling capability, they are still away from being able to detect very low power RF signal or capable to generate high RF power. Microwave circuits, therefore, are inserted to perform high-frequency processing so the ADC, and consequently the DSP unit, can more easily handle the RF signal. Such high-frequency processing functions are: amplification (low-noise, gain or high-power), signal combining or splitting, signal routing, phase and amplitude modulation, signal measurement, finally frequency generation and conversion.

A possible way to cluster these functions is divide them in front-end or back-end functionalities. The prior are typically connected to the radiating element, implementing low-noise or high-power amplification and some form of signal routing and phase and amplitude modulation. Instead, microwave back-ends are connected to the ADC and therefore provide all the functionalities so the RF signal can be profitably delivered to the digital section. Such functions are typically more complex functions such as extraction of signal characteristics, frequency generation and conversion.

The two sub-system and the relevant circuits will be discussed in the following.

In this chapter, we will not describe signal filtering, being this an extremely extensive topic, excellently covered by Matthaei et al. work [1] and other chapters in this book.

2. Front-end systems and their circuits

As briefly described in Section 1, a microwave front-end system is directly connected to the radiating element. In most cases, these subsystems provide dual mode operation: receive and transmit mode. In receive mode, the incoming RF signal is very weak and its power needs to be amplified to an adequate level, without adding excessive noise content or distorting the RF signal.

In transmit mode, the outgoing RF signal has to be raised to the highest possible value in order to guarantee an adequate transmission level. In this case too, distortion must be limited in order to preserve the information carried by the RF signal.

Signal routing is often necessary to implement the desired RF path between the digital section and the selected antenna.

2.1 Antenna front end

Antenna front ends (AFE) are employed to condition the received signal coming from the antenna port to make it usable for the following sub-systems, an example of which is described in Section 3.

The main functions of an AFE circuit are: low-noise amplification, protection against strong interference, signal routing when multiple I/O ports are present, and partitioning into sub-bands if needed. **Figure 1** depicts a simplified schematic.

The first, leftmost, section of the AFE contains the protection and signal routing function. The protection against strong interfering signals is accomplished using a limiting circuit commonly realized through a shunt diode. The limiter has to be the first circuit in order to protect the following components from strong interference signals that might damage sensitive circuitry. Next, there is a signal routing section

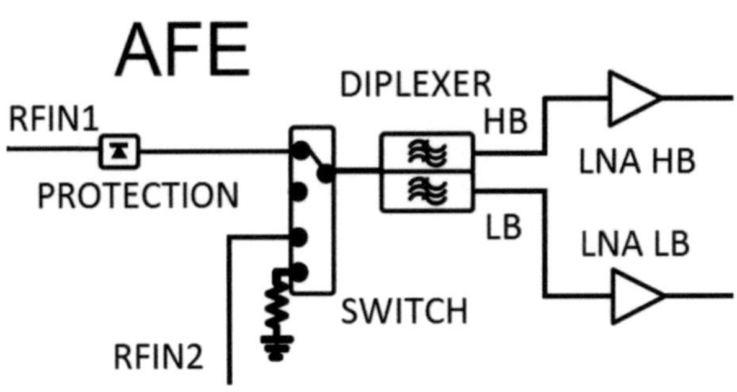

Figure 1.
Antenna front end (AFE) schematic diagram.

(switch), which is necessary when multiple inputs are present. Switching circuits are described in Section 2.1.2.

Once the signal has been routed, and strong interference has been eliminated then the signal is fed to a frequency diplexer (if necessary) and subsequently to a low-noise amplifying (LNA) stage. The frequency diplexer is inserted when the following stages operate at sub-bands that are less than the total RF input band-width (BW). In **Figure 1**, the bands are indicated as low-band (LB) and high-band (LB).

The single sub-bands can cover a decade BW, so that the overall BW of the module is more than a decade. Finally, the LNA is the key-component of AFE circuit and its role and properties are described in Section 2.1.1.

Figure 2 reports the AFE circuit's noise figure and gain in the two LB and NB sub-bands. This is a typical performance that can be accomplished by using COTS (commercial off-the-shelf) available components.

The gain (dashed lines) are plotted on the left axis while the NF (solid line) on the right. Some observations in the following: first, the effect of the input diplexer is quite evident around 2.5 GHz were the two gains cross. This is a side-effect of applying the RF at one input and then di-plexing into two sub-bands. The relatively high NF in LB and HB is mainly due to all the passive and protection circuits before the LNAs-Reasonably, all there passive structure will account for 4/5 dB losses. Additionally summing the NF of the LNA will lead to 5 dB in LB and typical 8/10 dB in HB. The gain ripple, more evident in the HB sub-band is due to the electrically long interconnects at microwave frequencies. Superior performance in terms of NF can be obtained by designing by oneself the critical circuits (i.e. LNA).

2.1.1 Low-noise amplifiers (LNA)

Low-noise amplifiers are an omnipresent component in any microwave receiving system. The LNA's role is to increase the power of the input signal, usually very low especially in long-distance communications, without adding an excessive noise contribution that would make the signal unmanageable by the following stages. The LNA's key characteristics are its gain (G) and noise figure (NF). Secondary, but still important parameters are linearity, power consumption and port matching.

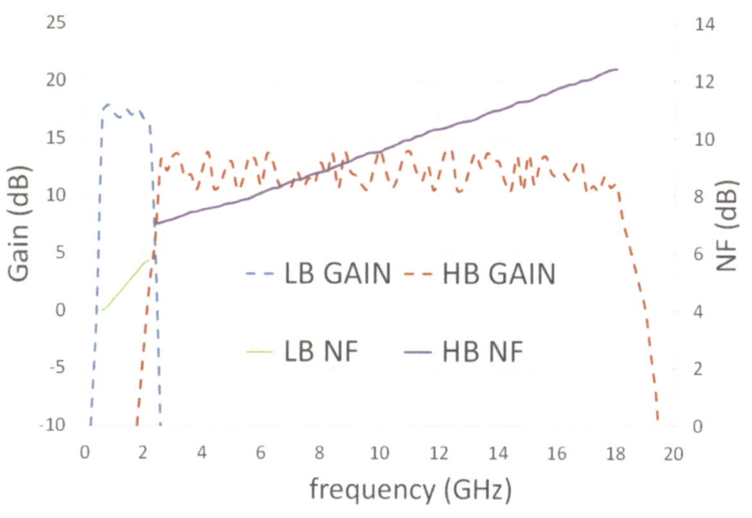

Figure 2.
Antenna front end (AFE) typical RF performance: gain (dashed) and NF (solid). Operating BW is 0.5–18 GHz.

All these parameters are influenced by on the maximum operating frequency, bandwidth and semiconductor technology. Understandably, performance tends to degrade as the frequency and bandwidth increase. Ideally, G should be high while NF should be the lowest possible. Below 10 GHz, 30 dB gain and less than 1.0 dB NF are suitable numbers, reachable when the circuits are realized in III–V compound semiconductors[2]. Above 10 GHz, some degradation has to be accepted in terms of greater NF and smaller gain.

The impact of the LNA's NF and G on system performance can be estimated using Friis' well-known formula that computes the system's cascade noise figure as a function of each stages' NF and G. An important consequence of this formula is that the overall NF of a radio receiver is primarily established by the NF of its first amplifying stage.

Subsequent stages have a weaker effect on signal-to-noise ratio. For this reason, the first stage amplifier in a receiver should be the LNA. Otherwise, as in **Figure 1**, the trade-off between NF and robustness (protection against strong interference) must be accepted.

Regarding the semiconductor, GaAs represents an interesting trade-off between performance and technology readiness level. GaN is slightly less performing, in terms of NF, but has the benefit of handling much more power, making it suitable in receivers where the presence of strong signals is foreseen.

An LNA must satisfy linear, noise, power and intermodulation requirements. Often linear and noise performance require opposite design choices, i.e. matching for noise or matching for gain. Simultaneously satisfying requirements often in contrast between them is not simple at all.

Luckily enough, many design strategies have been described, s dome of which dating back to the 1970s [3] up to more recent ones [4]. In the latter, a comprehensive design strategy that simultaneously accounts for linear and noise requirements is presented. Most of these strategies have a limited bandwidth since feedback is computed at the central design frequency. In [5] a survey of GaAs LNAs operating at very different frequencies (from 5 to more than 100 GHz) is presented. The most suitable design technique is indicated depending on the LNA's operating frequency.

On the other hand, there are other deign topologies, mainly distributed, that are capable of obtaining UWB performance. **Figure 3** depicts the circuit schematic and micro-photo of decade bandit LNA operating between 2 and 18 GHz [6].

The LNA depicted in **Figure 3**, demonstrates 23 dB typical gain and 4 dB typical NF over the entire 2–18 GHZ BW. Another interesting feature is its capability of withstanding high input power signals, demonstrated up to 10 W RF continuous wave. Gain and noise figure of the LNA reported in [6] is plotted in **Figure 4**,

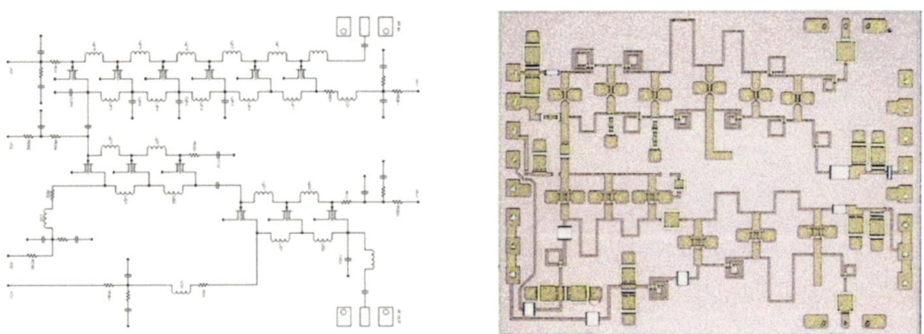

Figure 3.
Schematic circuit topology (left) and micro-photograph (right) of an UWB 2–18 GHz GaN UWB distributed LNA.

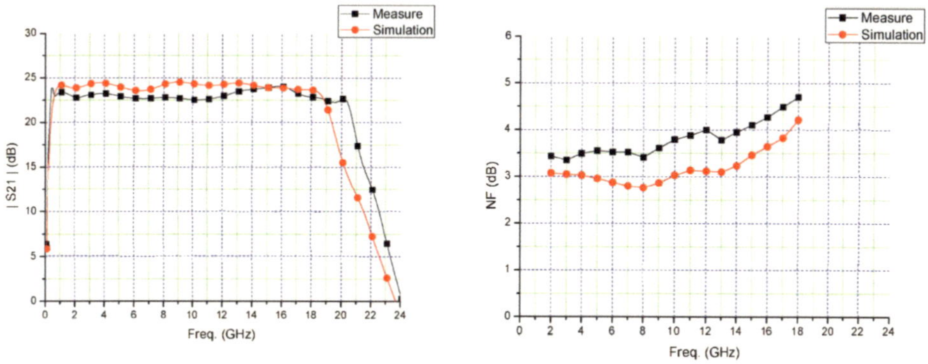

Figure 4.
UWB 2–18 GHz GaN distributed LNA simulated vs. measured gain (left) and NF (right).

demonstrating the LNA's capability to obtain more than 20 dB over a very wide operating BW. In the same condition, the NF averages at 3 dB.

2.1.2 Switching circuits

Switching circuits in microwave system are used to implement signal routing therefore performing path selection. Usually they have one common input port and N possible output ports, and only one can be selected at a certain instant.

The switching device can be either a diode or a Field Effect Transistor (FET). As usual, each possibility has its pros and cons, and will be discussed in the following. The diode switch has better loss performance; it can be fractions of dB even at tenths of GHz. On the contrary, FET switches are quite *lossy* and very easily reach 1–2 dB insertion loss even below 10 GHz. Apart from this very important parameter, all other aspects tend to be in favour of the FET.

Firstly, the FET is voltage controlled and does not dissipate any DC power thanks to the very high impedance of the gate terminal. On the contrary, diodes require a large current to achieve their low loss state, and therefore some DC power is dissipated across the diode. Secondly, the FET has faster switching time, i.e. the time required to select a different output once the appropriate external command has been received. The switching time in FET switches can be as low as a few nanoseconds. Diode switches may require tenths of nanoseconds to change their state since the direction of the bias current needs to be reversed, and this is not immediate considering the stray capacitances in the control section and the diode itself. Finally, FET switches are more robust and linear, especially if realized in wide band-gap semiconductors as GaN. They can tolerate up to tenths of Watts, while the switch diodes seldom survives incident powers above a few Watts.

Isolation, i.e. the unwanted leakage to an unselected path, is another important parameter in switches. However, this performance mainly depends on the selected switch topology, rather than the selected technology.

The frequency behaviour of the two technologies is comparable, especially when small gate length transistors are employed. In both cases, diodes and FETs, acceptable performance up to 50 GHz, and even beyond, are achievable.

Consequently, the choice between FET and diode switches, should be carried out considering losses, power handling and switching time requirements.

Several UWB switching topologies have been proposed and validated. Typically, an inductor is inserted in the switching circuit to resonate the diode's or FET's OFF-state parasitic capacitance [7]. A resistor is also inserted, in this way a more uniform behaviour is obtained over a larger operating BW. The schematic applying this technique is depicted in **Figure 5** (left) together with its physical implementation (right). The two compensating inductors and resistors are labelled with L_1/ R_1 and L_2/R_2, while the two compensated FETs are Q_1 and Q_2.

The UWB switch shows an insertion loss lower than 2.2 dB, an isolation higher than 25 dB, and a power handling capability better than 38.5 dBm at the 1 dB compression point in the entire bandwidth. The SPDT's key performance is plotted in **Figure 6**.

GaN-HEMT technology therefore demonstrates a good level of maturity for microwave power switch applications and as such is becoming the reference technology for specific high-end applications.

2.2 Transmit/receive modules

Transmit/receive modules (TRM) are the key building block of most telecommunication apparatus, Radars and many other Electronic Systems. Their role is to process the RF signal in both operating modes: transmit and receive. Typically, a TRM operates in a half-duplex manner, i.e. in a certain instant it is either in receive mode or in transmit mode, therefore processing either a received signal or a signal to be transmitted. A possible schematic diagram of a T/R module is reported in **Figure 7**; a TRM is always connected in some way to a radiating element (Ant).

In order to keep the dimension of the TRM as small as possible, some circuits are involved in both transmit and receive mode and therefore need to process the signal independently form the port at which it arrives. Referring to **Figure 7**, such components are the switches (SWT), the attenuator (ATN) and the phase shifter (PHS). Incidentally, the latter is required in systems that perform beam steering and can be avoided elsewhere. The attenuator instead can be inserted for multiple purposes: it can be used to prevent strong RF signals leaking to the following circuits or to obtain beam amplitude tailoring, in phased arrays.

Critical components are the LNA, already described, and the high-power amplifier (HPA) described in the following section. Another critical component is the switch connected to the antenna port. The key feature of this element is to show very low losses. High losses would entail an unacceptable degradation of both received and transmitted signal. In the past, for high frequency applications, this element was often a bulky ferrite circulator. With the advent of GaN semiconductor, well-known for its superior power handling capabilities, MMIC technology has become the standard. Finally, the gain control section in the receive path is used to attenuate strong incoming signals. It is seldom used in transmit mode, since in most application the goal is to transmit as much RF power as possible.

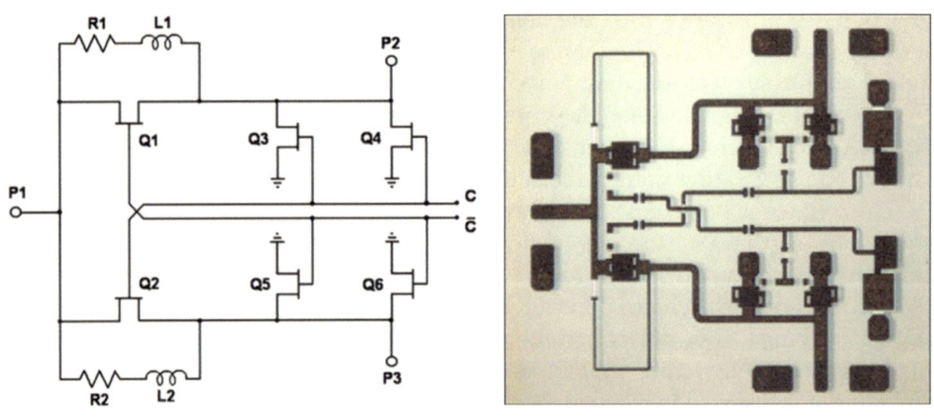

Figure 5.
Schematic circuit topology (left) and micro-photograph (right) of an UWB 2–18 GHz GaN switch.

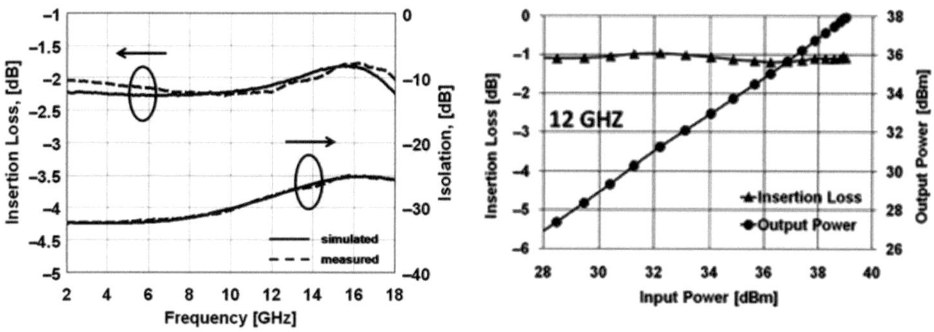

Figure 6.
UWB 2–18 GHz GaN switch measured insertion loss and isolation over the full operating BW (left) and non-linear performance at 12 GHz (right).

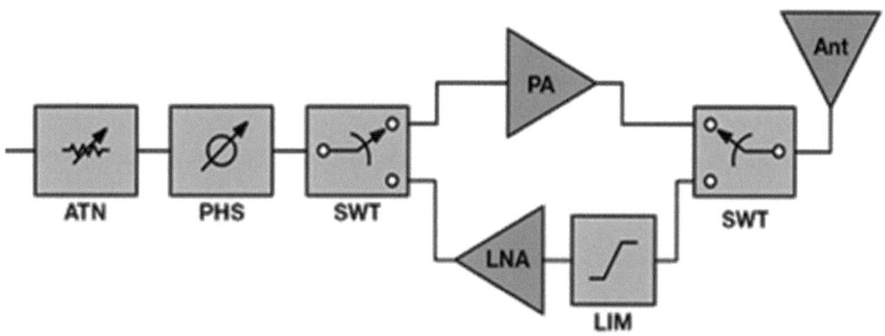

Figure 7.
Schematic diagram of a TRM.

In Electronic Warfare systems, which notoriously manage UWB signal to contrast different emitters, the maximum to minimum operating frequency of the TRM can be as high as 6-to-1. Higher ratios became unfeasible since the increase in the BW would be obtained to detriment of performance, and, in any case, it could be impractical since it is very challenging to design directional antennas having wider bandwidths anyway.

For other aerospace and avionics applications, the operating BW is typically 20–40% the value of the central operating frequency.

Figure 8 reports the key-parameters of a highly-integrated GaAs-based compact TRM for EW [8].

The typical output power is around 5 W (37 dBm) while the RX gain is on average 20 dB. Such performance was accomplished through Multifunction chips and ASIC component integration in new multi-layers technology (Roger 4003/Cu/FR4) were adopted in order to reduce cost, space, production life cycle and increase integration level. High output power in the transmit mode was achieved using a 4 W wideband amplifier and by minimizing circulator loss.

2.2.1 High-power amplifiers (HPA)

The HPA is the key component of any microwave transmission systems, and its performance may have a huge impact on the final system architecture.

Its role is to boost the transmitted signal's power without adding undesirable signals generated by distortion. At the same time, the HPA should be efficient, in terms of its capability of transforming to the power provided by the DC supply RF power. From a technology point of view, they come in at least two variants: vacuum

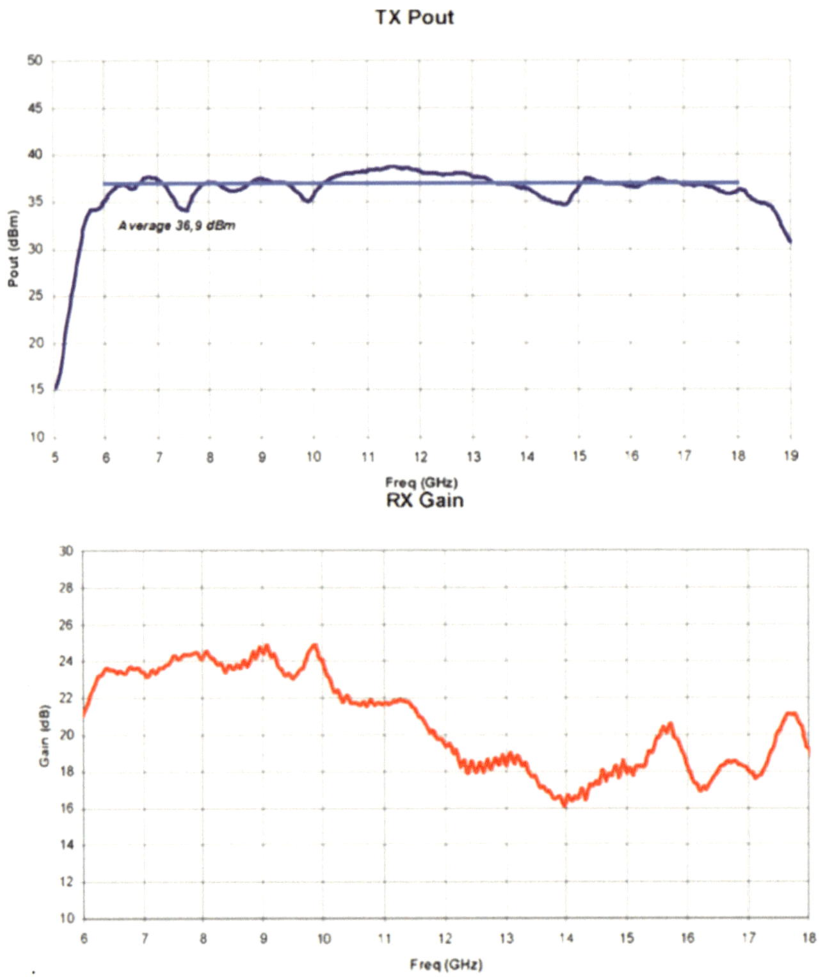

Figure 8.
Key performance of an UWB 6–18 GHz GaAs TRM output power in TX (top) and gain in RX (bottom).

tubes and solid state circuits. With the advance of semiconductor technologies, vacuum tubes are becoming a legacy product. Nonetheless, they still provide a valuable solution when the power to be transmitted is in the order of tenths of kilowatts. Typically this requirement is related with a few avionic and spaceborne applications. The advantages of solid state device in terms of ruggedness, size, reliability, performance and cost are such that, whenever a solid state alternative becomes accessible, it quickly becomes adopted by the System Engineering team.

Since its first appearance in R&D labs at the beginning of the new millennium, GaN has travelled a long way, and has now become the standard semiconductor, even in ADS systems, where reliability and process repeatability is a main concern. The advantages of GaN, over other III–V semiconductor, for high-power and high-frequency systems, reside on its capability to deliver a high amount of RF power in a small footprint, with little or none thermal management issues. Especially the last feature, make GaN attractive for ADS applications, often operating in harsh thermos-mechanical environments.

MMIC GaN HPAs are capable of delivering hundreds of Watts at low microwave frequency (<5 GHz), tenths of watts at microwave frequencies (5–20 GHz), and some Watts even at millimetre-wave.

Scientific literature focusing on the design of HPAs is practically infinite, and here we will give an extremely short hint on some design topologies. Usually HPA are synthesized by combing, at the HPAs output, the power provided by some FETs (where some is usually a power of 2). Moreover, suitable techniques can be applied to increase the output power and efficiency [9]. These techniques rely on synthesizing the output impedance to respect an optimum condition both at the operating frequency and also at its higher order harmonics. As you can imagine this can be complicated, especially in wideband applications.

If a very large BW is sought, then other circuit topologies become handy. For example, distributed amplification is well known for its UWB frequency response. Simultaneously, a *cascode* transistor topology can be applied [10], increasing even more the amplifier's BW.

3. Back-end systems and their circuits

As briefly described in Section 1, a microwave back-end system is responsible for delivering the RF signal—or better its information content—to the ADC and consequently the DSP unit or to the low-frequency (often referred to as VIDEO) analogue stages.

Typically, this is accomplished through frequency conversion, when the DSP performs A/D sampling, or by performing some manipulation on the RF signal so its power and/or frequency component can be determined by the subsequent stages.

3.1 Frequency conversion

UWB downconverters and up-converters usually require multi-stage conversion plan since a single frequency conversion would not be able to eliminate all over spurs or leakage of the Local Oscillator (LO) signal. In fact, a very large sweep of LO frequency would be needed to down-convert the required portion of the large input RF BW into the smaller Intermediate Frequency (IF) BW. Therefore, at some point, there will be inevitably a strong intermodulation product or harmonic of LO that would fall in the IF BW. To overcome this issue a multi-stage frequency conversion topology, depicted in **Figure 9**, is advisable when the RF BW is large.

Here we will discuss in detail the down-converter architecture, but similar assumptions and design goals hold for the up-converter.

The first stage of the schematic depicted in **Figure 9**, is the filter bank, so the UWB signal RF IN signal is split into smaller adjacent sub-bands. Typically, each sub-band is less than an octave, and consequently the number of filters depends on

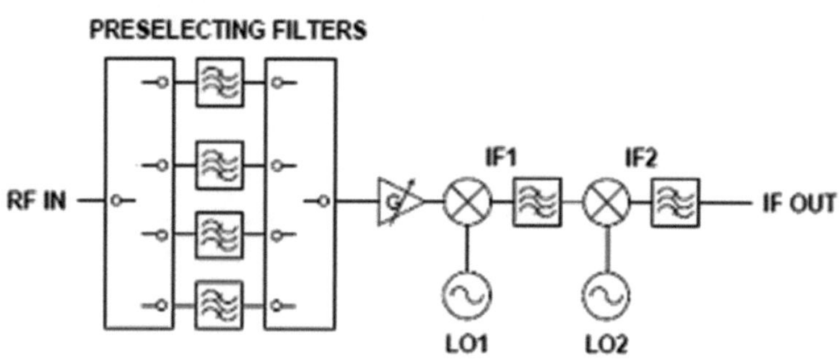

Figure 9.
Multi-stage down-converter topology.

the ratio of the maximum to minimum RF IN frequency. In this way, the input RF signal is preselected ensuring that only the required portion of the input RF BW is fed to the following stages.

The next section consists in a frequency translation, performed by a set of mixers. Even in a downconverter system, often this first section of mixing performs an up-conversion. This is to shift the spurs to a very high frequency, well above the IF1 range. The value of LO1 is inversely proportional to the selected RF frequency. In this way, the sum of incoming RF IN signal and LO is constant and equal to IF1. Keep in mind that, the first stage is typically an up-converter even if the overall sub-system performs down-conversion. Some variable gain is also inserted in the RF chain to level out the gain response. In fact, as the input RF frequency increases, the RF losses become more evident and need to be compensated by lowering the attenuation accordingly.

At this point, all the desired signal falls within IF1, whose BW is much smaller than the UWB signal. RF IN, and can now be more easily down-converted to IF2 which is the sub-system's output frequency. A single LO2 frequency is sufficient in most cases, while in other UWB applications a variable LO2 could be required. In some extreme cases, for example when the RF BW is larger than a decade, a triple frequency conversion could be required [10].

This topology is definitively more complex than the single down-converter case, and requires a high frequency IF1 which often could be in the millimetre wavelength. Undoubtedly, such system complexity is the price to pay for having a UWB down-converter with high spurious free dynamic range (i.e. negligible spurs or LO harmonics falling in the IF OUT BW).

A key performance indicator of a frequency-converter is the Spurious-Free Dynamic-Range (SFDR). This parameter quantifies the ratio (expressed in dB) between the fundamental (desired tone) and the intermodulation product having the highest power inside the IF band. SFDR can be seen as an indicator of the down-converter's capability to perform its characteristic function without injecting unwanted signals at the output IF. Such unwanted signals, referred to as intermodulation components (defined in the following Section 3.1.1) can be mistaken by the receiver as low power real signals, but in reality they are an unwanted by-product of the real signal's down-conversion.

Intermodulation products are unavoidable, however the important matter is that they are below a given threshold therefore becoming undetectable and will not produce 'false signals' at system level. SFDR usually has a characteristic behaviour vs. frequency since, referring to **Figure 9**, LO1 and LO2 change in order to select different input RF bands, and therefore will produce different intermodulation orders when down-converting different RF input bands. **Figure 10** depicts the SFDR of an UWB 2–18 GHz downconverter [11] where a three Local Oscillators architecture is employed.

3.1.1 UWB mixers

Mixer circuits are employed to translate the information applied to the RF carrier to a different frequency, namely IF, more easily processed by other circuits, typically of a digital nature. Given the incoming RF signal, and the LO signal, the frequency components at output of the mixer are:

$$\text{Output spectrum} = |m * RF + n * LO| \qquad (1)$$

where m and n are positive or negative or null integers. In a down-converter the desired Intermediate Frequency (IF) value is obtained when $m = 1$ and $n = -1$ or vice versa. In an up converter, *IF* appears instead of *RF* in the previous relation and $m = n = 1$ applies.

LO frequency is selected to trade-off between feasible *LO* values and harmful intermodulation products falling close to *IF* frequency. Undesirable intermodulation components are obtained when $m \neq 1$ or $n \neq 1$. The order of the intermodulation product is defined as: $order = m + n$. Usually the power level decreases as the order increases.

UWB mixers are typically realized through a double balanced, or double-double balanced (sometimes referred to as a triple) topology to eliminate the most annoying intermodulation and LO harmonics. Spurs and harmonic rejection is accomplished by appropriately combining the mixing signal through hybrid quadrature couplers (BALUNs or similar circuits). Ultimately, the UWB behaviour is limited by the coupling structures since the mixing device's behaviour can be considered ideal. UWB mixers often came in a connectorized package as the example depicted in **Figure 11**. It is also possible to obtain UWB Mixers in MMIC technology, however it is more difficult to realize UWB combiner/splitters in the confined dimensions required by MMICs.

An extensive description on the of mixers and their property can be found in Dr. Maas's comprehensive study: Microwave Mixers [12].

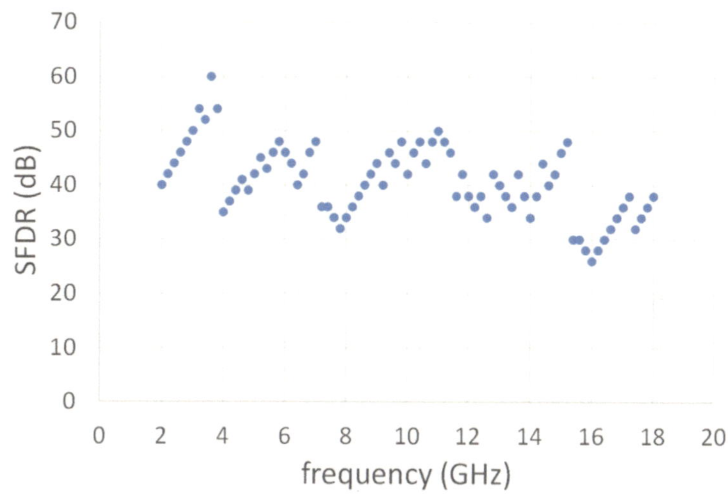

Figure 10.
Multi-stage down-converter SFDR vs. frequency.

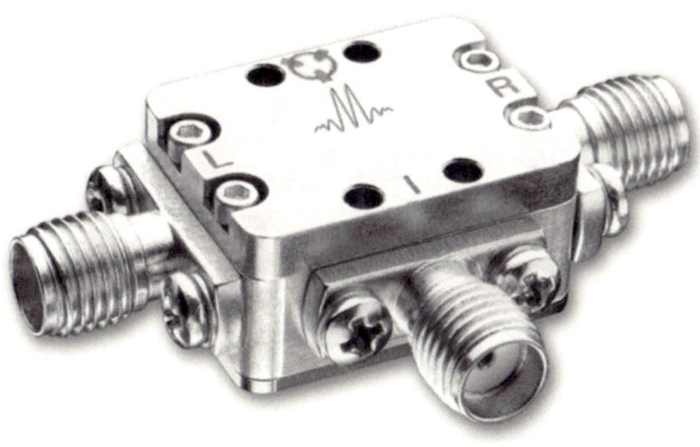

Figure 11.
Example of connectorized mixer (photo courtesy of Marki Micorwave).

3.2 Microwave measurements

In some applications, the incoming RF signal is unknown and therefore the system requires to quickly determine some of its key features, for example, carrier frequency and envelope power. Performing these measurements on UWB RF signals using only Digital HW can be complicated, or in some cases even impossible considering the power and frequency limitations of Silicon based technology. For this reason, microwave circuits have to be inserted in the receiving system to perform preliminary yet fast signal characterization.

3.2.1 RF power measurement

The power measurement of an RF signal is commonly performed by appropriately feeding the RF signal to a diode—or to an array of diodes.

The diode has a well-known input-to-output square law characteristic, around the origin, when the diode is slightly forward biased. The signal coming out from the diode is composed by the even order harmonics of the incident RF signal and, most of all, its zero order term that is a DC voltage. The latter is captured by the following stages to perform the power measurement. There are several circuits that are capable of performing such functions, and their topology depends on the overall BW of the RF signal and also its dynamic range, i.e. the maximum to minimum power ratio to be analysed. A simplified block diagram of the diode detector is shown in **Figure 12**, together with the waveforms at each section of the circuit.

A detector logarithmic video amplifier (DLVA) consists of a diode detector followed by a logarithmic video amplifier (LVA), and its simplified schematic is given in **Figure 13**.

As explained before, the diode (detector circuit) converts RF signal into a DC voltage which is then fed to an amplifier with a logarithmic transfer function. Let us analyse the role of the two circuits.

The diode's quadratic law provides the DC voltage component of the incoming RF signal. The level of such DC voltage can vary significantly: from tenths of μV to some Volts. The ratio between the maximum and minimum voltage could be unacceptable for the subsequent processing stages. A LVA is therefore inserted to compress the dynamic and make it more usable for the following stages. In fact, the log amplifier will greatly amplify the weak signals leaving the strong signal practically unaffected. It is worthwhile noting LVA's operating BW is a few tenths of MHz

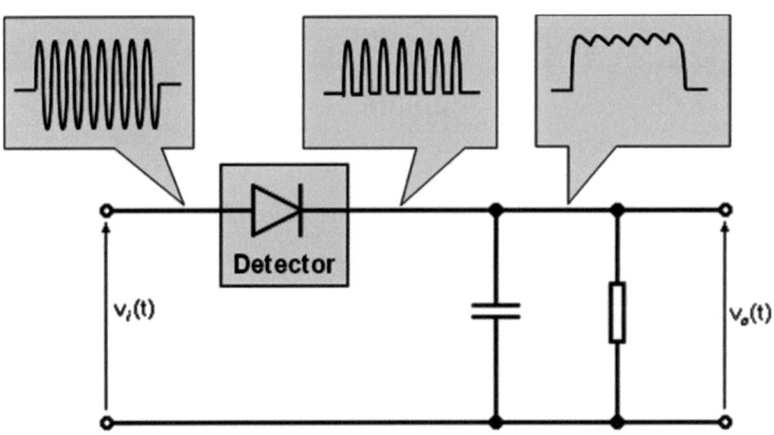

Figure 12.
Simplified schematic of a diode detector and waveforms at each node of the circuit.

as opposed to the incoming signal which may be a few tenths of GHz. The dotted line in **Figure 14** depicts the linear relationship between input power in dBm and output power in Volts when using a LVA after the diode. The same **Figure 14** reports also the I/O characteristic without LVA, solid line.

It appears that the log relationship is very useful for performing power measurements, since its gradient is constant (approximately 25 mV/dB in **Figure 14**). The response without LVA is practically null up to −5 dBm and then rises instantaneously to +1.0 V after −5 dBm, making it unpractical. When dynamic range is not critical, a DLVA with a 30- or 40-dB dynamic range may provide sufficient performance to help capture and process all signals present.

When the dynamic range of the incoming RF signal is greater than the one accepted by the diode, then the schematic depicted in **Figure 15** can be used to increase the system's dynamic range.

Basically, the input signal is split in two paths: one path having high RF gain and one path having low RF gain hence an RF attenuated path. Then, the VIDEO signal coming out from both DLVAs is summed at the output.

The principle of operation is the following: when the signal is low, only the amplified DLVA detects thanks to the RF gain (G) before the DLVA. When the signal power level becomes high, the gain path is practically saturated, delivering a constant voltage value, while the attenuated (Att) path performs the additional power measurement whose voltage is increased by the constant term coming from the saturated gain path.

Finally, it is worthwhile noting that the BW of these circuits is practically limited by the BW of the RF circuits preceding the diode. In fact, the latter is often capable

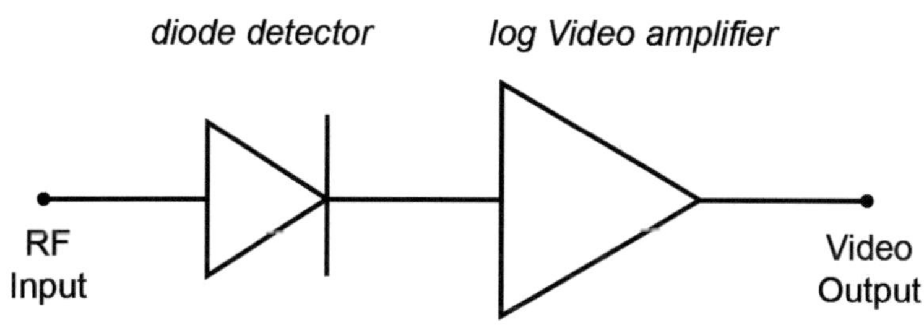

Figure 13.
DLVA simplified schematic.

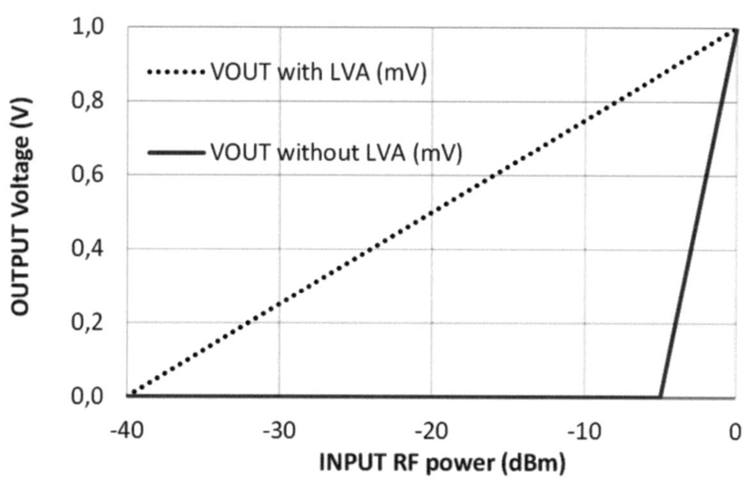

Figure 14.
Output voltage (with and without LVA) as function of input power expressed in dBm.

of processing RF signals up to tenths of GHz. Diode operating at 50 GHz are not uncommon.

Another way to increase the system's dynamic range is to use a SDLVA topology, schematically depicted in **Figure 16**.

An SDLVA (Successive Detection Logarithmic Amplifier) is similar to a DLVA however, the SDLVA circuit is designed in such a way that it does not need a detector before the logarithmic video amplifier. The SDLVA uses multiple compressive stages of RF gain to emulate the exponential transfer function. The output of each stage is coupled into a linear detector. The detector operates over a narrower dynamic range, which means that more detectors are needed to cover the same dynamic range.

The principle of operation is the following: when the RF signal power is low, all amplifiers operate in a linear condition (i.e. the output power is proportional to the input power) and consequently the DC voltage provided by the diode detector is proportional too. As the power increases, the final stages begin to saturate and their output is capped to a saturation value. Therefore, any additional output voltage will be delivered only from the first stages until they saturate too, saturating as a consequence the Video output at its maximum values.

The typical dynamic range of each detector is approximately 10 dB, which are then summed in a single video amplifier so as to provide a single detected output. The overall dynamic range is $10 \times N$ dB where N is the number of amplifiers. $N = 7\text{–}8$ represents an acceptable trade-off between high dynamic range and circuit feasibility.

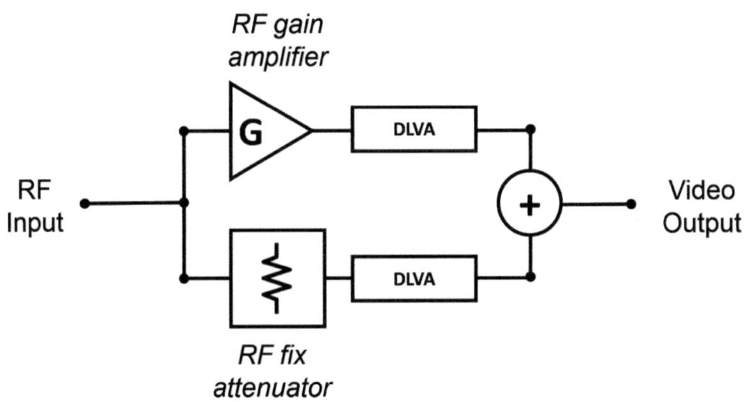

Figure 15.
Simplified schematic of an extended DYNAMIC range DLVA.

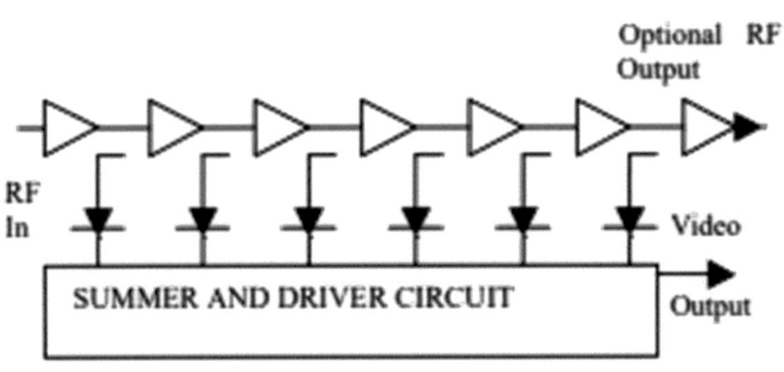

Figure 16.
SDLVA simplified schematic.

3.2.2 RF frequency measurement

Microwave circuits can be profitably employed to perform an estimation of the carrier frequency of the incoming RF signal. This is all but a simple task. Nowadays, the advance of digital components has made feasible digital frequency measurement up to 10 GHz and beyond. Even in this case, some form of microwave front-end is required to prepare the signal for digital sampling. Above tenths of GHz, and considering UWB signals, pure digital frequency estimation becomes unpractical and microwave circuits have to be inserted.

UWB microwave circuits that perform frequency estimation are referred to as Instantaneous Frequency Measurement (IFM), whose very simplified schematic diagram is reported in **Figure 17**.

The RF input $v_{in}(t) = A\ cos(\omega t)$, where $\omega = 2f$ and f is the instantaneous frequency carrier, is hard-limit amplified and then split by a power divider in two equal amplitude signals, one of which is delayed with respect to the other through a delay line. Hard limitation consist in amplifying all incoming signal to a fixed power level—obviously within a feasible dynamic range. The role of the delay line is to out-phase the two signals coming from the power divider's outputs by a quantity proportional to the frequency carrier.

The higher the frequency the more the two signals applied to the mixer will be out-phased.

Considering the relationship IF = $|m^*RF + n^*LO|$ already introduced in Section 3.1.1, we have that the valuable mixer's output will be a zero frequency product and therefore a DC voltage since *RF* and *LO* signals have, of course, the same carrier frequency. The amplitude of the DC voltage is proportional to cosine of the out-phase between the two signals and therefore to the carrier frequency, trough the relationship:

$$Vout(t) = K\ cos(\Delta\Phi), \tag{2}$$

and

$$\Delta\Phi = 2\pi * \Delta L * f/v_p \tag{3}$$

where ΔL is the difference between the physical distance of the delay line and the direct pat; f is the carrier frequency, while v_p the speed of the EM wave in the medium. Higher order mixing terms are eliminated by the low-pass filter (LPF). The BW of such circuit is limited by the BW the power divider and the mixer. As a consequence, the delay is synthesized to implement π shift (maximum out-phase) at the components maximum operating frequency.

Very often Vout is digitized by means of a N-BIT analog-to-digital converter (ADC). This helps the subsequent stages since the information is provided digitally, but the information accuracy is limited by the number of BITs (typically not more than four) and quantization effects. To overcome this issue the scheme in **Figure 18** is applied.

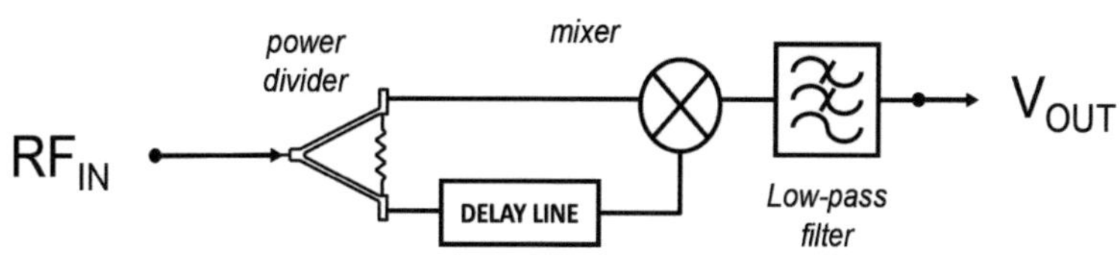

Figure 17.
IFM simplified schematic.

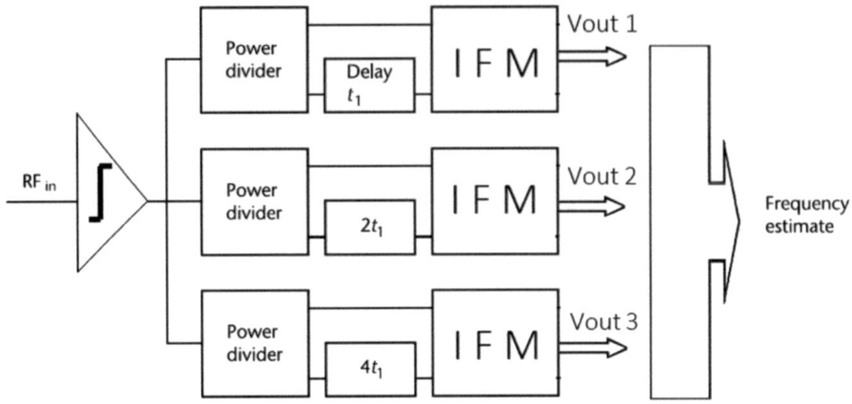

Figure 18.
Functional block diagram of a THREE-base harmonic IFM receiver.

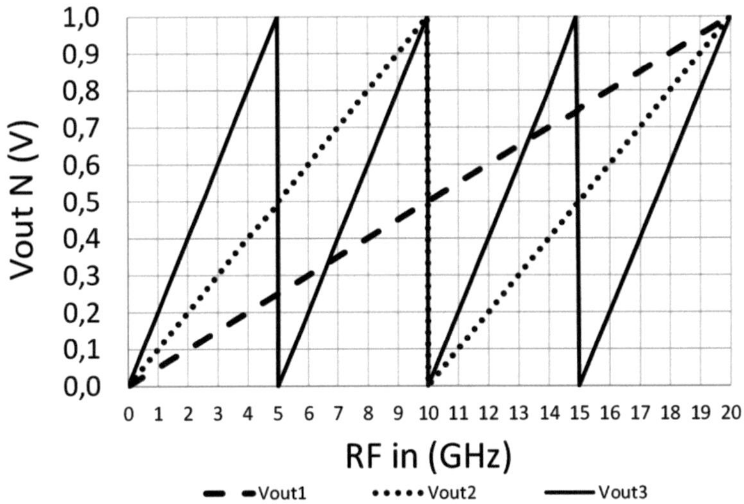

Figure 19.
Operational characteristics of a THREE-base harmonic IFM receiver.

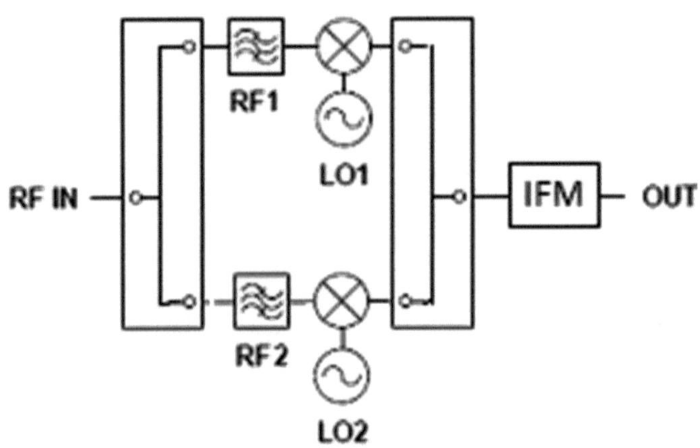

Figure 20.
UWB frequency measurement circuit simplified schematic.

The input signal is split into *N* lines (three in the example depicted in **Figure 18**), the three signals are the delayed by appropriately dimensioned transmission liens. Each line has double the electrical length of the preceding section. Delay t1 is determined so that the signal measuring is unambiguous at the maximum operating frequency (worst case for ambiguity discrimination. In practice, the shortest time delay (t1) is used to resolve the frequency measurement ambiguities while the largest time delays (t2 = 2*t1 and t3 = 4*t1) provide a fine frequency measurement accuracy $\Delta f3 = \Delta f1/8$.

The three output voltages (Vout 1, 2 and 3) values vs. input frequency are depicted in **Figure 19**. Vout 1 (dashed line) is the only transfer function that is unambiguous vs. the input frequency. Therefore, this value is used to obtain a coarse yet unambiguous information of the input frequency. The other two voltages (Vout 2 and 3) suffer from ambiguity, but their resolution (in terms of $\Delta V/\Delta f$) is better and therefore provides an ambiguous yet accurate information. The real frequency estimation is performed by correlating the three output voltages values.

If a larger BW is sought, then the topology depicted in **Figure 20** becomes useful.

The RF input signal is split in two or more paths. The number of parallel output branches depends on the BW of RF IN; a larger BW requires a greater number of paths. Once the signal has been divided, it is filtered into sub-bands and then translated, through mixing, at the frequency of operation of the IFM. The sum of $RFn + LOn$ must be constant for each branch and equal to the operating frequency of the IFM. In order to determine the actual RF IN frequency, a detector—not reported in **Figure 20**—must be inserted in each branch so the system can discriminate between the various possibilities. The branch in which the detector reads power corresponds to the sub-band of interest. In theory, the BW can be extended by increasing the number of adjacent sub-bands. There are practical limitations, and the number of sub-bands is seldom greater than four.

4. Multi-functional integrated circuits

The trend to integrate several function in one MMICs is unavoidable given the low SWaP-C constraints of modern electronic systems for high-end applications. At the moment it is possible to realize a full T/R module using only two MMICs, each one realized in the appropriate semiconductor therefore leveraging its benefits and peculiarities [13], and possibly in the near future there will be an all-on-one multi-functional RF MMIC. **Figure 21** depicts a transmit/receive module diagram, very similar to the one described in Section 2.2, where the area delimited by dotted lines indicate the two multifunctional MMICs that fulfil all TRM functionalities. The two MMICs are the Single Chip Front End (SCFE) and the Core-Chip (CC). The prior is directly connected to the antenna, while the latter to the microwave back-end and the DSP sections. Red and blue arrows indicate the direction of Rx and Tx signal. In some cases, black lines, the signal travels in both direction according to transmit or receive mode.

4.1 Single-chip front-end (SCFE)

As described in Section 2.2, RF TRMs are usually realized by interconnecting several functionalities fabricated on separate MMICs. Such an approach is somehow the result of the inability of a single compound semiconductor technology to properly carry out the main features normally required by an RF TRM. In some applications, such as active electronically scanned array (AESA), this multichip approach could result in a suboptimal overall system. Indeed, the density of the radiating

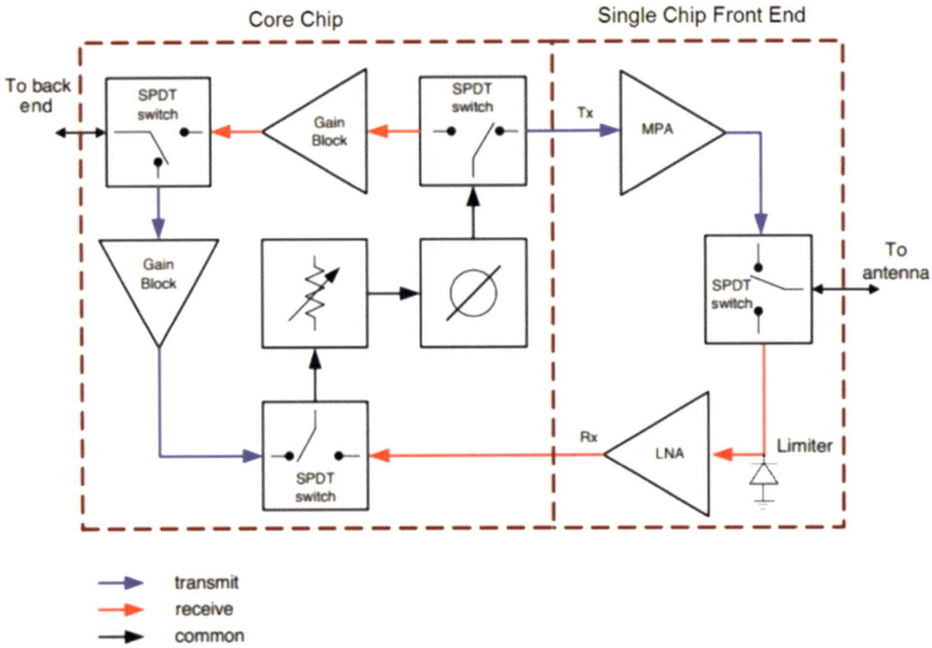

Figure 21.
Two MMIC compact TRM.

elements, and thus the performance of the array, is directly related to the physical size of the overall TRM that is adopted behind each elementary radiating element. Thus, the availability of highly integrated front ends with sensible smaller foot print and weight could be very useful for such systems, since it would allow the increase of radiating elements density, leading to the performance maximization.

Recently, with the advance of highly performing and reliable commercial GaN processes, the concept of single-chip front end (SCFE) was investigated as a possible alternative to multichip TRMs.

Essentially, an SCFE is a single MMIC that integrates the three functionalities that are required for a half-duplex TRM, i.e. the HPA, the LNA, and the single-pole double-throw (SPDT) switch. GaN offers remarkable advantages in terms of reliability, robustness, heat dissipation and power handling capability, as compared to GaAs counterparts for this specific application.

The limiter inserted before the LNA, in **Figure 21**, is optional, and depends on the maxim incident power and the LNA's robustness.

Figure 22 depicts the schematic block diagram and implementation of a recently published SCFE [14].

4.2 Core chip (CC)

The same driver that pushes the advancement of SCFE circuits, generally referred to as low SWaP-C requirements, is behind the development of core-chips (CC) too.

As opposed to GaN used in SCFE, CCs are developed in GaAs thanks to its superior high frequency performance, and most of all the possibility to employ simultaneously enhancement and depletion mode transistor, which are the necessary for the digital logic on board CCs.

Essentially, a CC is a single MMIC that integrates the three functionalities that are required for signal conditioning and routing in modern TRMs, i.e. the phase shifter, the attenuator, and the single-pole double-throw (SPDT) switches used to select the transmit and receive mode operations. All these functionalities require a large number of control signals to set the state of the attenuator, the phase shifter and the TX or RX mode. Typically these states are set through a dozen or more separate controls and it is unfeasible to deliver them in parallel way (i.e. one interface pad for each control signal provided to the CC). A work-around for this problem consists in sending the control signal serially, and performing de-serialization on board. Such function requires circuits as Flip-Flop, latches, buffers and level shifters. Such circuits have a purely digital function, and are synthesized, as stated before, enhancement and depletion mode transistor, which are in Silicon based technology—having unfortunately a severe RF power and frequency limitation- and also GaAs that on the contrary is very suitable for high frequency RF signal conditioning.

Core chips come in, at least, two architectural variants. In the first, referred to separated architecture [15], the RF signal travels bi-directionally in the phase shifter and attenuator as depicted in **Figure 23a**. In the common-leg architecture, **Figure 23b**, the signal travels always in the same direction in both transmit and receive modes. More switches are required, at least three as opposed to the single T/R switch in separated architecture, to implements the correct routing of the single in Transmit and receive mode. Architecture (a) has the advantage of being more simple and compact, while architecture (b), although more complex, has the advantage of having amplifiers in the common leg (being mono-directional) therefore improving noise figure, gain and linearity performance.

The physical implementation of a CC operating at X-band (9–11 GHz) is reported in **Figure 24** [14].

Finally, let us focus on the role of the Serial-to-Parallel converter (S2PC). At least 13 lines (commands) are necessary to set the state of the 6-BIT phase shifter ($\Delta\Phi$), 6-BIT Attenuator (ΔA) and SPDT switch for T/R mode setting. If the S2PC were not on board then 13 lines would have to be fed to the CC. With the insertion of the S2PC only one serial data line is necessary together with a clock line and an Enable command. Therefore only three command lines, plus a digital voltage supply line, as opposed to 13 or more. The S2PC is the very dense area depicted in **Figure 24**.

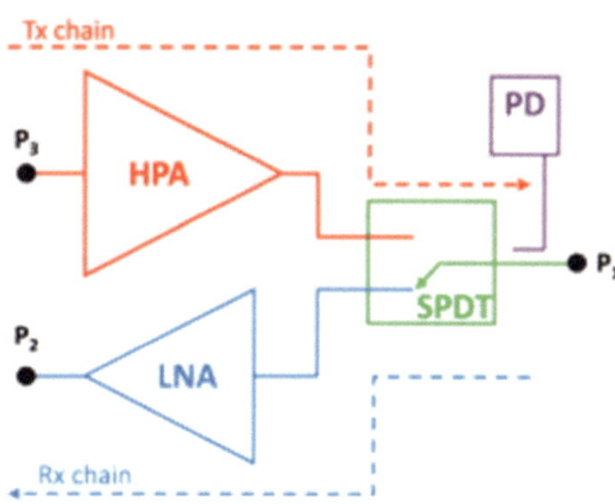

(a)

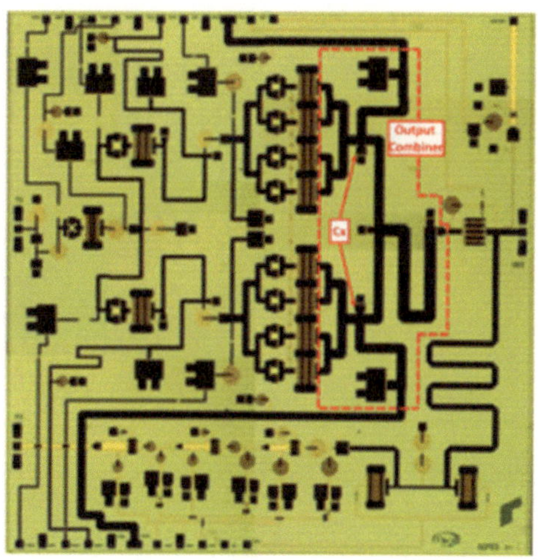

(b)

Figure 22.
(a) Block diagram and (b) photograph of a realized SCFE operating around 22 GHz. Size is 49 mm².

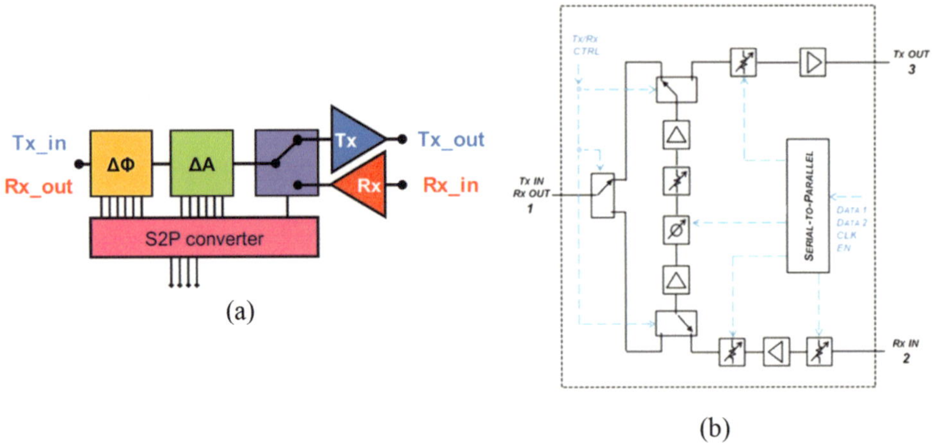

Figure 23.
Block diagram of separated architecture CC (a) and common-leg architecture (b).

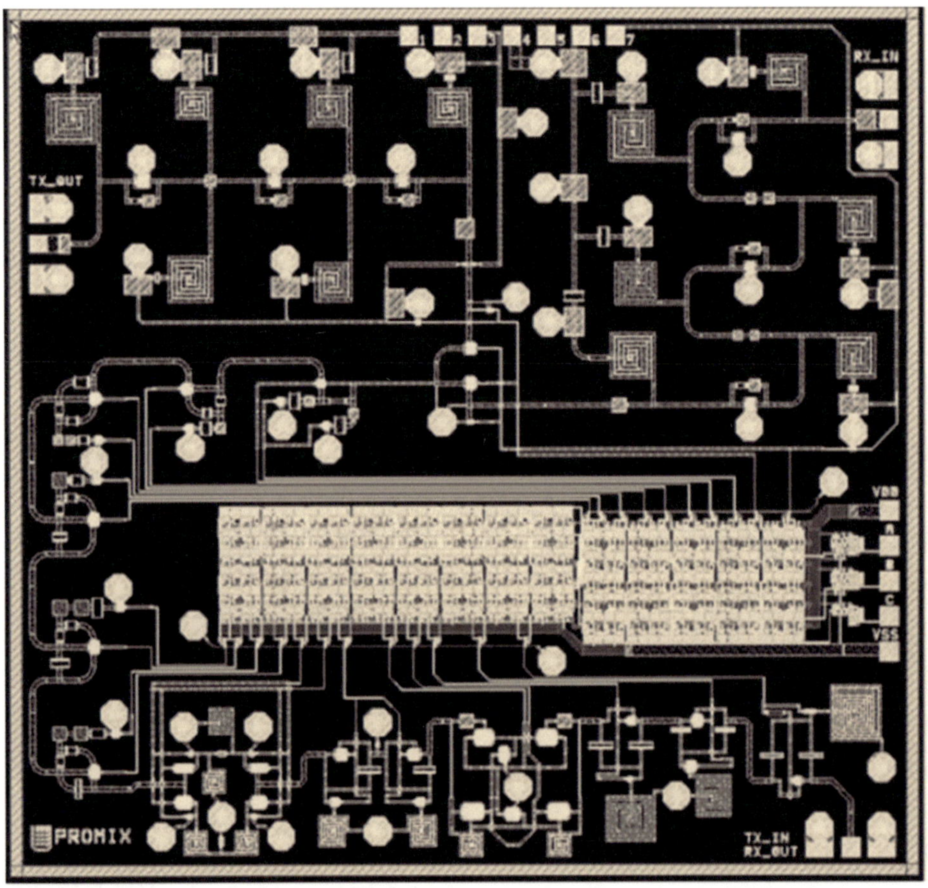

Figure 24.
Photograph of a realized CC operating around 10 GHz. Size is 15 mm^2.

5. Conclusions

UWB microwave circuits play a vital role in modern electronic systems for Aerospace, defence and Security applications. They are inserted to appropriately transfer the RF signal to/from the radiating element and perform some preliminary signal processing: phase and amplitude control, frequency conversion, and measurement of some fundamental signal characteristic.

Topologies and technologies need to be carefully leveraged in order to obtain the best possible performance. While in the past the trend was to design one circuit to implement one functionality applying the *divide-et-impera* (divide-and-conquer) paradigm nowadays we are looking at *all-in-one* highly integrated solutions. Of course this integration comes to the expanse of much higher circuit synthesis and analysis complexity.

Acknowledgements

Patrick E. Longhi would like to thank friends and colleagues at Elettronica Group in Rome (ITA) for the many fruitful and insightful technical discussions during his time spent in the company as a Microwave Engineer.

Author details

Ernesto Limiti and Patrick E. Longhi*
Electronic Engineering Department, University of Rome Tor Vergata, Rome, Italy

*Address all correspondence to: longhi@ing.uniroma2.it

References

[1] Matthaei G, EMT J, Young L. Microwave Filters, Impedance-Matching Networks, and Coupling Structures. Chichester: Wiley; 2011. 1132 p. DOI: 10.1002/9780470974704

[2] Niehenke EC. The evolution of low noise devices and amplifiers. In: IEEE/MTT-S International Microwave Symposium Digest. DOI: 10.1109/MWSYM.2012.6258248

[3] Lehmann RE, Heston DD. X-band monolithic series feedback LNA. In: IEEE MTT-S International Microwave Symposium Digest; June 1985; DOI: 10.1109/MWSYM.1985.1131900

[4] Ciccognani W, Longhi PE, Colangeli S, Limiti E. Constant mismatch circles and application to low-noise microwave amplifier design. IEEE Transactions on Microwave Theory and Techniques. 2013;**61**(12):4154-4167

[5] Ciccognani W, Limiti E, Longhi PE, Renvoisè M. MMIC LNAs for radioastronomy applications using advanced industrial 70 nm metamorphic technology. IEEE Journal of Solid-State Circuits. 2010;**45**(10):2008-2015

[6] Limiti E et al. An ultra-broadband robust LNA for defence applications in AIGaN/GaN technology. In: IEEE MTT-S International Microwave Symposium Digest; 2010. pp. 493-496

[7] Ferrari M et al. High-power monolithic AlGaN/GaN high electron mobility transistor switches. International Journal of Microwave and Wireless Technologies. 2009;**1**(4):339-345. DOI: 10.1017/S1759078709990183

[8] Baccello D, D'Antoni M, Orobello B, Sperduti E. Miniaturised low cost solid state 4W TXRX common leg for 6-18 GHz phased array. In: 2010 IEEE MTT-S International Microwave Symposium; pp. 401-403

[9] Colantonio P, Giannini F, Limiti E. High Efficiency RF and Microwave Solid State Power Amplifiers. John Wiley & Sons, Ltd; 2009. DOI: 10.1002/9780470746547

[10] Kobayashi KW. Broadband design techniques and technology for future wireless and wire-line applications. In: Asia-Pacific Microwave Conference; December 2010

[11] Longhi P, et al. Compact wideband downconverter module for electronic warfare applications. In: European Radar Conference (EuRAD); October 2009

[12] Maas S. Microwave Mixers. 2nd ed. Norwood, MA: Artech House; ISBN: 9780890066058

[13] Bentini A et al. A C-Ku band, 8 channel T/R module for EW systems. In: IEEE International Conference on Microwaves, Antennas, Communications and Electronic Systems (COMCAS); November 2017; DOI: 10.1109/COMCAS.2017.8244719

[14] Giofrè R et al. S-band GaN single-chip front end for active electronically scanned array with 40-W output power and 1.75-dB noise figure. IEEE Transactions on Microwave Theory and Techniques. 2018;**66**(12):5696-5707. DOI: 10.1109/TMTT.2018.2861716

[15] Longhi P et al. A compact high performance X-band core-chip with on board serial-to-parallel conversion. In: Proceedings European Microwave Conference; September 2010; pp. 902-905

Chapter 3

Inner Tapered Tree-Shaped Ultra-Wideband Fractal Antenna with Polarization Diversity

Sarthak Singhal and Amit Kumar Singh

Abstract

A coplanar waveguide (CPW)-fed third iteration inner tapered tree-shaped ultra-wideband (UWB) fractal antenna for polarization diversity applications is presented. The antenna comprises of two orthogonal fractal antenna structures to achieve polarization diversity performance across the frequency spectrum of 4.7–19.4 GHz. An isolation of more than 15 dB is accomplished. The designed antenna has a nearly omnidirectional radiation pattern with an average gain of 2.45 dB, very low values of envelope correlation coefficient and capacity loss, nearly constant diversity gain (DG) and mean effective gain (MEG) values. The time domain analysis results illustrated the low dispersion in the radiated pulse. The designed antenna has advantages of wider bandwidth and miniaturized dimensions along with good diversity performance. These advantages make the designed antenna a promising candidate for future wireless communication systems having multipath fading as a major concern.

Keywords: coplanar waveguide feeding, fractal antenna, polarization diversity antenna, ultra-wideband antenna

1. Introduction

In 2002, FCC allocated the unlicensed frequency spectrum from 3.1 to 10.6 GHz for ultra-wideband (UWB) technology [1]. After this allocation, ultra-wideband has received attention from wireless communication experts owing to its advantageous features like wider bandwidth, low cost, low susceptibility to multipath fading, reduced probability of detection and intercept and potentially high data rates. In a highly dense and dynamic environment, the UWB systems suffer from multipath fading due to reflection and diffraction. This multipath fading results into the degradation of signal-to-noise ratio (SNR) and channel capacity.

An effective method to resolve these multipath fading issues is the incorporation of antenna diversity techniques in wireless communication systems. Several types of diversity, such as space/spatial, pattern and polarization diversity, have been already proposed and implemented to receive multiple signals [2–4].

In a diversity scheme, the power or signal-to-noise ratio of the received signal is optimized by the selection or combining of output signals in several ways like selection combining, equal gain combining or maximal ratio combining. The detailed description of diversity combining techniques is available in [5, 6].

For good diversity performance, the received signals should have very low correlation between them [7]. An increase in correlation reduces the combining efficiency. In a spatial diversity scheme, a large separation (compared to wavelength) between the antennas is used to achieve decoupling between signals. This large space requirement limits the use of this diversity method. To overcome this drawback, other techniques such as pattern or polarization diversity [8, 9] are investigated. These alternate techniques involve the use of two or more antenna elements with different radiation patterns [10]. An UWB system with polarization diversity technique has potential applications in advanced instruments used for microwave imaging, radar and high-speed data transfer. Some UWB polarization diversity antennas are already reported in the literature [11–28]. However, the application of those available structures is limited due to their large dimensions, multilayer structure, complex feedline, complex geometries, etc.

Among the various bandwidth enhancement techniques, the use of fractal geometries is proven to be a good method. Fractal antenna structures have a compact size and wideband performance due to properties of self-similarity, space filling and effective energy coupling properties [29].

In this chapter, a compact CPW-fed UWB fractal antenna with polarization diversity performance is presented. The bandwidth of the antenna structure [29] is enhanced by loading the coplanar ground planes with a quarter wavelength long rectangular notches. Two identical copies of this antenna structure are arranged orthogonally to achieve good interport isolation and orthogonal polarization diversity performance without affecting the UWB performance. In the following sections, antenna design description is followed by discussion of frequency domain analysis results, time domain analysis results and diversity performance parameter calculation. Finally, it is concluded with major findings of this chapter.

2. Antenna design

The geometry of the antenna structure is demonstrated in **Figure 1**, and its optimized dimensions are listed in **Table 1**. It is etched on a 1.6 mm thick FR-4

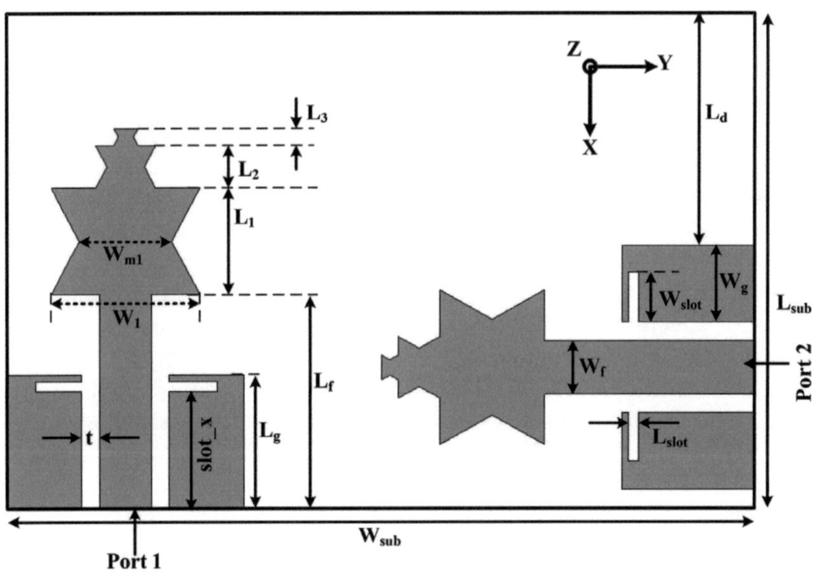

Figure 1.
Geometry of the antenna.

Dimension	Value (in mm)	Dimension	Value (in mm)
L_{sub}	18.5	W_{sub}	28.7
L_g	5	W_g	2.9
L_f	8	W_f	2
L_1	4	W_1	5.8
L_2	1.6	W_{m1}	3.6
L_3	0.64	L_d	8.65
slot_x	4.35	t	0.7
L_{slot}	0.4	W_{slot}	1.85

Table 1.
Dimensions of the designed polarization diversity antenna.

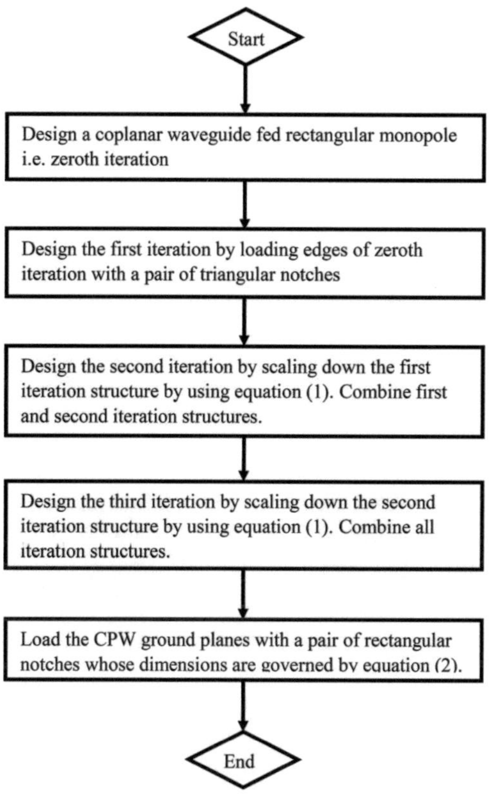

Figure 2.
Flow chart of designing initial radiator geometry.

epoxy substrate having a relative permittivity of 4.4 and loss tangent of 0.02. All the dimensions are optimized by using finite element method (FEM)-based Ansoft's high-frequency structure simulator (HFSS) [30]. During simulation, the radiating patch and ground planes are assumed to be perfect electrical conductors. The antenna structure is designed in two steps.

The fourth iterative fractal geometry of radiator is derived from a rectangular monopole by loading it with a pair of triangular notches on its edges. The flow chart of designing the intermediate design steps is presented in **Figure 2**. The intermediate design steps for radiator geometry are shown in **Figure 3**. The iteration structure dimensions are governed by Eq. (1):

$$R_n = R_{n-1} r^n \tag{1}$$

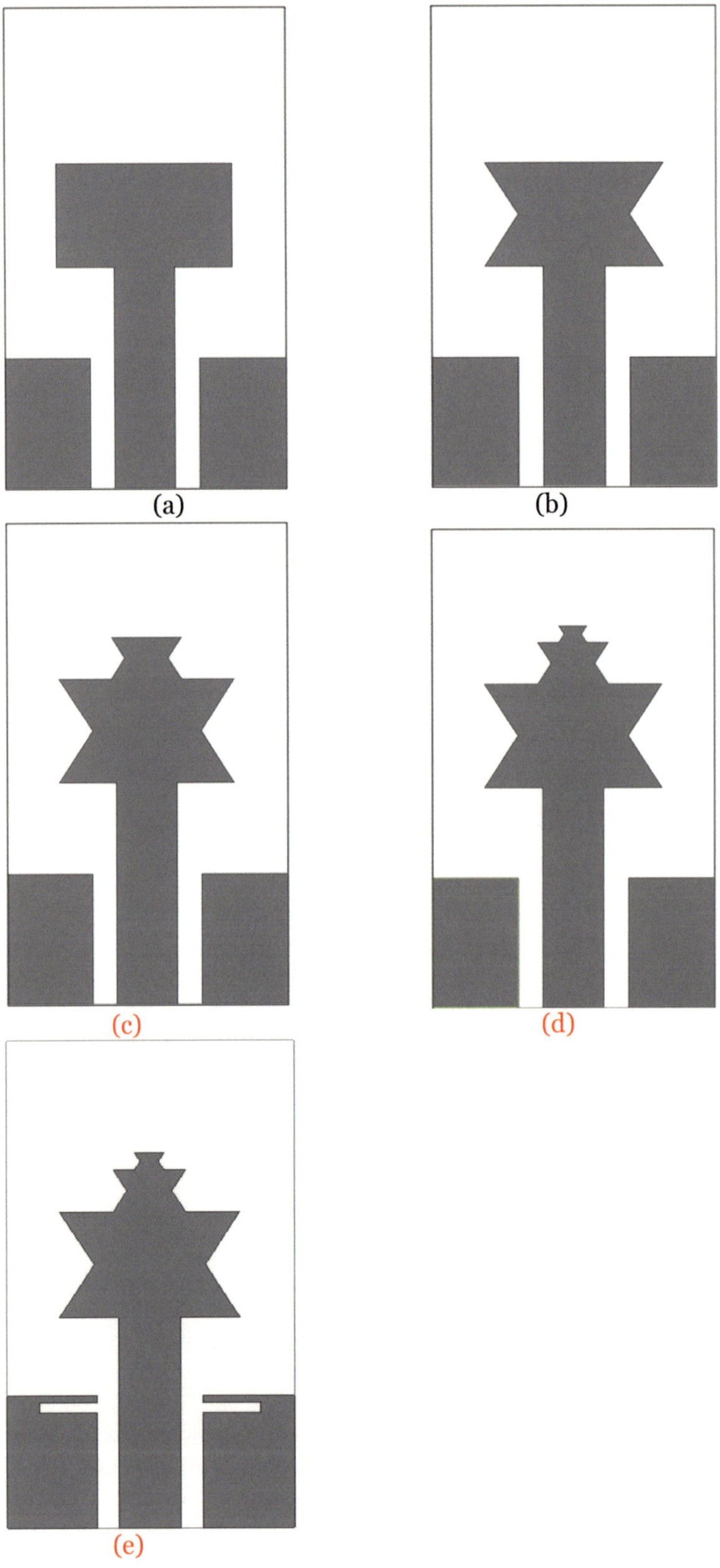

Figure 3.
Designing stages of proposed MIMO antenna element (a) Zeroth iteration, (b) First Iteration (c) Second Iteration (d) Third Iteration (e) Third iteration with ground notch.

DOI: http://dx.doi.org/10.5772/intechopen.86071

where:
n = iteration number = 2 or 3.
R_1 = dimension of the first iteration, i.e. L_1, W_1 and W_{m1}.
R_n = dimension of the nth iteration.
r = iterative ratio = 0.4.

In the first step, the dimensions of radiating patch are unaltered. The bandwidth of inner tapered tree-shaped fractal antenna is enhanced by loading the ground plane with quarter wavelength rectangular notches to excite an additional resonance at 18.6 GHz. The dimensions of notch are governed by Eq. (2):

$$L_r = \frac{c}{4f_r} \approx L_{slot} + 2 \times W_{slot} \tag{2}$$

L_r = electrical length of slot for resonance.
f_r = resonance frequency = 18.6 GHz.

In the second step, two identical structures designed in the first step are placed orthogonal to each other. The air gap between the two structures and their locations are optimized. All the frequency domain results are calculated by using HFSS. The time domain results and diversity performance parameters are analysed by using CST MWS [31].

3. Results and discussion

The intermediate antenna design steps are compared in terms of their S_{11} curves in **Figure 4**. It is observed that the bandwidth is increasing with an increase in iteration. For further increase in iteration, no significant improvement is observed. The reflection coefficient curves for the initial antenna structure with and without ground notches are illustrated in **Figure 5**. Its quantitative analysis is listed in **Table 2**. It is observed that the lower band edge frequency is negligibly changed, whereas the higher band edge frequency is shifted from 16.4 GHz to 19.4 GHz in the case of notch-loaded ground plane. The initial resonances are slightly shifted to higher frequency with an additional resonance.

3.1 Frequency domain

The designed diversity antenna structure is simulated by using HFSS and CST MWS simulators. The variations of simulated scattering parameters with frequency are demonstrated in **Figure 6**. The quantitative analyses of bandwidth for two antenna elements used in designed antenna structure are presented in **Table 3**. From **Figure 6** and **Table 3**, it is observed that there are some discrepancies among the two simulation results. These discrepancies can be attributed to the different mesh size suitable for numerical techniques on which the simulators are designed. In addition to mesh size, it is also important to mention that in CST MWS the structure can be solved in single pass instead of solution for different frequency spectrum, i.e. 1–2, 2–4, 4–8, 8–16 and 16–32 GHz in HFSS. The differences between the S_{11} and S_{22} characteristics are due to asymmetrical structure with respect to substrate. A good isolation of more than 15 dB is achieved. The designed antenna has resonances at the frequencies of 6, 8, 10.8, 15.8 and 18.8 GHz.

The comparison among the designed antenna and previously reported polarization/pattern diversity antenna structures is listed in **Table 4**. It is observed that the designed antenna has wider bandwidth, good isolation and miniaturized dimensions than other structures.

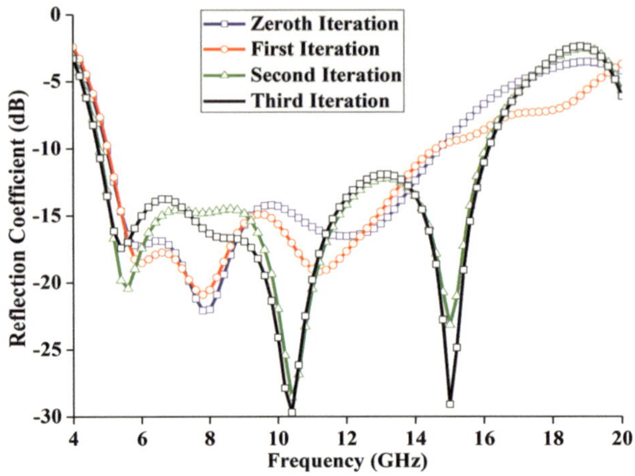

Figure 4.
Reflection coefficient of intermediate antenna design steps.

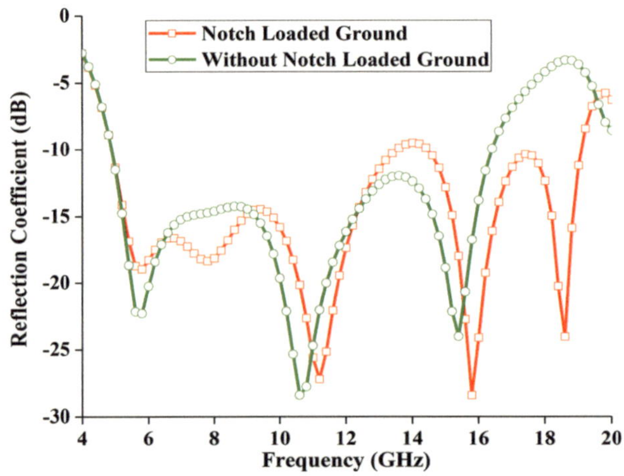

Figure 5.
Reflection coefficient versus frequency characteristic of single antenna element.

Antenna	f_L (GHz)	f_H (GHz)	BW (GHz)
Without ground notch	4.9	16.4	11.5
With ground notch	4.8	19.4	14.6

Table 2.
Bandwidth comparison of single antenna element with and without ground notch.

The simulated radiation patterns of two antenna elements at the resonance frequencies of 6, 8, 10.8, 15.8 and 18.8 GHz in all three planes are illustrated in **Figure 7**. From **Figure 7**, it is observed that the antenna structures have bidirectional and omnidirectional patterns at lower frequencies. At higher frequencies, the patterns are distorted omnidirectional in nature due to the excitation of higher-order modes at those frequencies. It is also clearly observable that the patterns of both antenna structures have a phase difference of 90° in each plane as desirable.

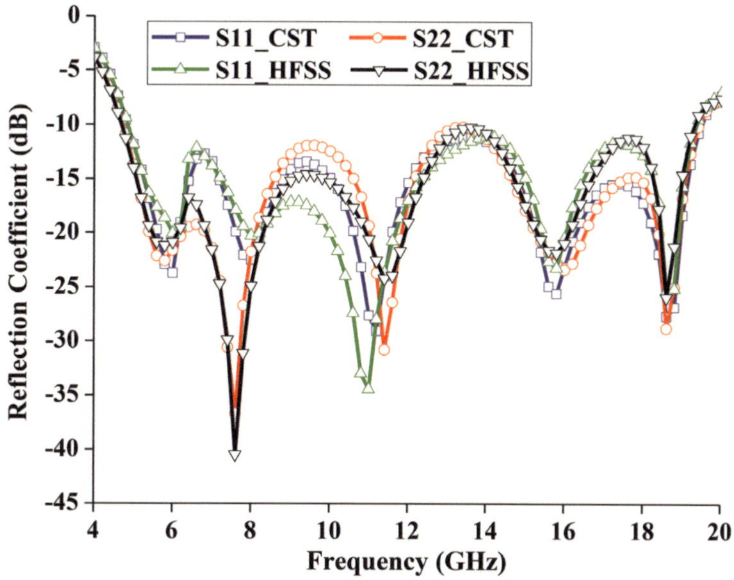

(a)

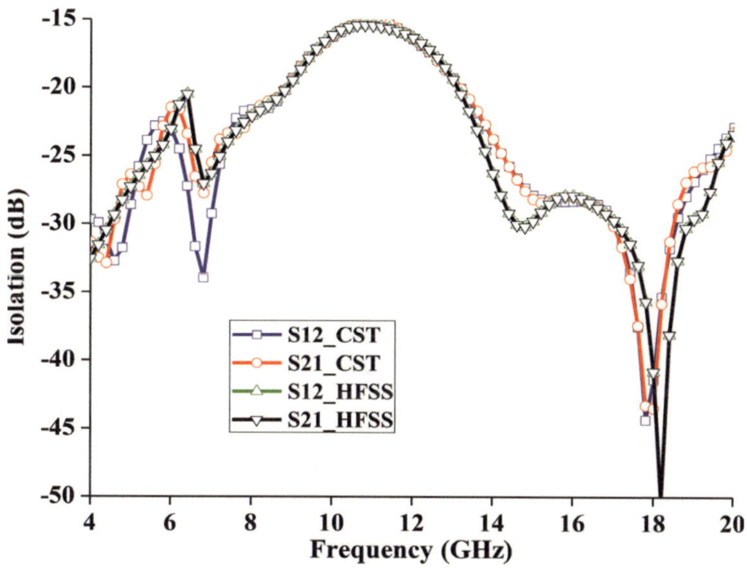

(b)

Figure 6.
Scattering parameters versus frequency characteristics of the designed antenna.

Figure 8 illustrates that the peak realized gain of Antenna I is varying from 0.52 to 4.98 dB with an average of 3.5 dB over the operating frequency band. It also presents that the gain of Antenna II is varying between 1.43 and 3.5 dB with an average of 2.45 dB.

The variations of radiation and total efficiencies with frequency for both antenna elements are shown in **Figure 9**. The radiation efficiencies of both antenna elements are more than 0.7 with an average of more than 0.86. Similarly, the total efficiencies have an average of 0.83. The efficiencies are decreasing with an increase in frequency due to the use of lossy FR-4 substrate.

	Method	f_L (GHz)	f_H(GHz)	BW(GHz)	%BW
Port 1	HFSS	4.85	19.4	14.55	120
	CST MWS	4.84	19.48	14.64	120.4
Port 2	HFSS	4.7	19.3	14.6	121.67
	CST MWS	4.7	19.4	14.7	121.99

Table 3.
Bandwidth of two ports.

S. No.	Chapter	Dimensions (mm^2)	Bandwidth (GHz)	Isolation (dB)	% BW	% size reduction
1.	[11]	80×80 mm^2	3-12	>15	120	91.7
2.	[14]	34×49 mm^2	3.1-10.6	>20	109	68.13
3.	[15]	58×58 mm^2	2.8–11	>14	75	84.22
4.	[16]	27×52 mm^2	3.25–12	>20	73	62.18
5.	[17]	45×50 mm^2	2.7-10.9	>25	121	76.4
6.	[18]	50×50 mm^2	2.76-10.75	>15	118	78.76
7.	[19]	23×39.8 mm^2	2.5–12	>21	79	42
8.	[20]	80×80 mm^2	3-12	>15	120	91.7
9.	[21]	42×81 mm^2	3.2-11.2	>20	111	84.39
10.	[22]	21×35 mm^2	2.9-11	>20	117	27.76
11.	[23]	24×26 mm^2	3.06-11	>18	113	14.91
12.	[24]	28×50 mm^2	2.8-11.5	>18	122	62.08
13.	[25]	30×50 mm^2	2.5-14.5	>20	141	64.6
14.	[26]	50×85 mm^2	2-9.5	>20	130	87.51
15.	[27]	42×30 mm^2	3.1-10.6	>20	109	57.86
16.	[28]	16×37.6 mm^2	3-12	>19	120	11.74
17.	**Proposed Antenna**	**18.5×28.7 mm^2**	**4.7-19.4**	**>15**	**122**	-

Table 4.
Comparison of designed antenna with other UWB diversity antenna.

3.2 Time domain

During time domain analysis, two identical antenna structures are arranged in two major configurations, i.e. face to face and side by side as shown in **Figure 10**. In each of these configurations, one port is excited, whereas the other port is terminated with a matched load. The antenna structure is excited with a Gaussian pulse having a centre frequency of 13 GHz and bandwidth of 1–25 GHz. The normalized amplitudes of the transmitted and received signals are presented in **Figure 11**. From these normalized amplitudes, the correlation between the two signals, i.e. system fidelity factor, is calculated by using Eq. (3). The calculated values of system fidelity factor for four cases are listed in **Table 5**. The values listed in **Table 5** indicate that the signal is slightly distorted in side-by-side configuration in comparison to face-to-face configuration for both cases.

$$F = max\left[\frac{\int_{-\infty}^{\infty} s_t(t)s_r(t+\tau)d\tau}{\int_{-\infty}^{\infty} |s_t(t)|^2 dt \int_{-\infty}^{\infty} |s_r(t)|^2 dt}\right] \quad (3)$$

where $s_t(t)$ and $s_r(t)$ are the transmitted and received pulses and τ is the group delay. The variations of group delay with respect to frequency for all four cases are illustrated in **Figure 12**. It is observed that the group delay has its variations less than 1 ns over the entire band of operation.

3.3 Diversity performance

To analyse the diversity performance of the designed antenna, various parameters like envelope correlation coefficient, diversity gain (DG) and mean effective gain (MEG) are to be calculated from s-parameters or farfield patterns. The envelope correlation coefficient (ECC) signifies the correlation between the radiation patterns of two antenna elements. For the designed antenna structure, ECC (ρ_e) is

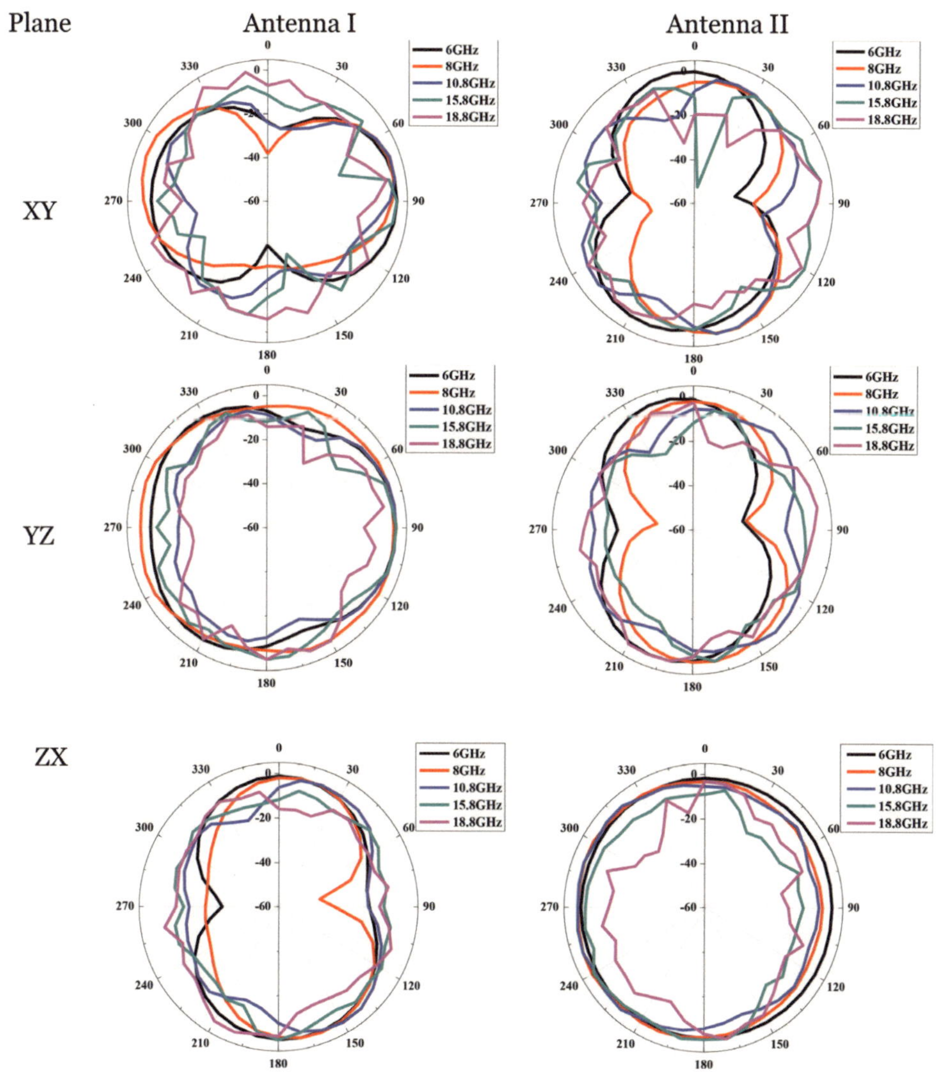

Figure 7.
Radiation patterns of the two antenna elements in XY, YZ and ZX planes.

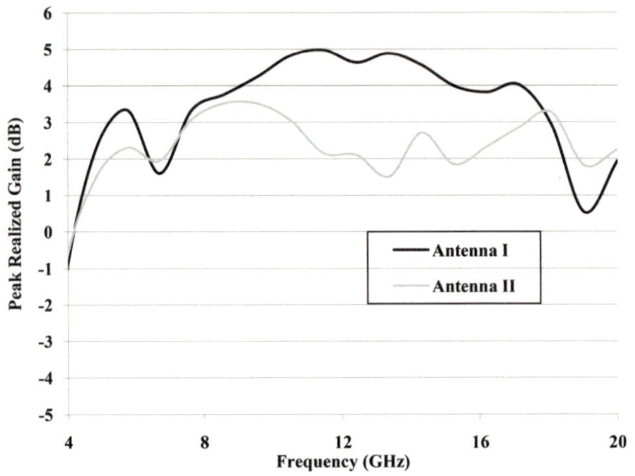

Figure 8.
Peak realized gain versus frequency characteristics.

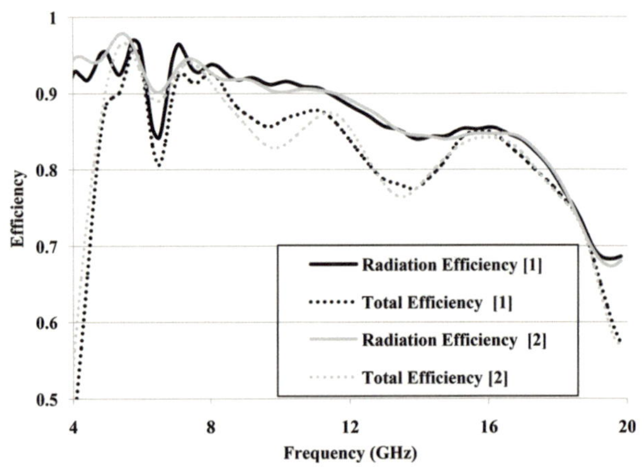

Figure 9.
Radiation and total efficiencies of the two antenna elements.

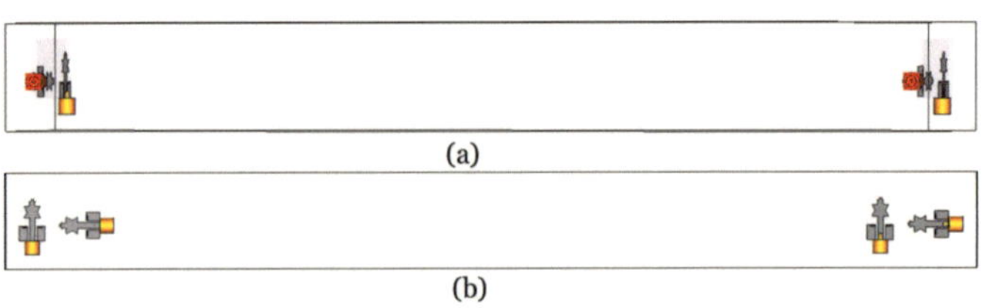

Figure 10.
Time domain analysis configurations of the diversity antenna. (a) Face to face (b) Side by side.

calculated by using Eq. (4) [32]. The calculated values of ECC are plotted in **Figure 13**.

$$\rho_e = \frac{\left|S_{11}^* S_{12} + S_{21}^* S_{22}\right|^2}{\left(1 - \left(|S_{11}|^2 + |S_{21}|^2\right)\right)\left(1 - \left(|S_{22}|^2 + |S_{12}|^2\right)\right)} \tag{4}$$

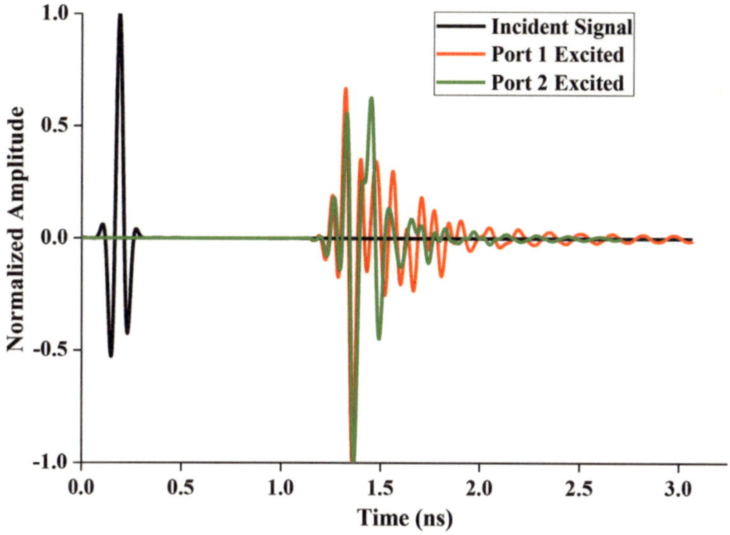

(a)

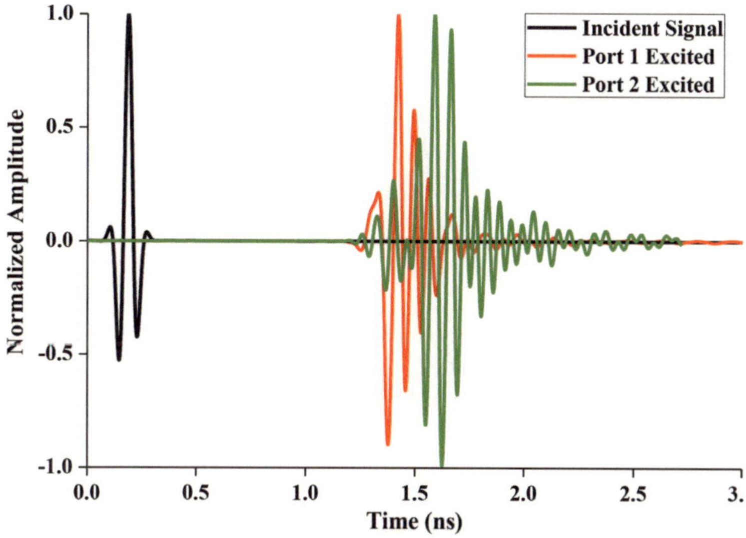

(b)

Figure 11.
Normalized amplitudes of the transmitted and received pulses in all four configurations. (a) Face to face (b) Side by side.

System Fidelity Factor	F2F	S2S
Port 2excited	0.79	0.67
Port 1excited	0.88	0.83

Table 5.
System fidelity factor for four configurations of the designed antenna.

From **Figure 13**, it is observed that the ECC values are less than 0.005 in the entire band of operation. These low values of ECC (<0.5) signify that the designed antenna is a good candidate for the UWB applications with polarization diversity [16].

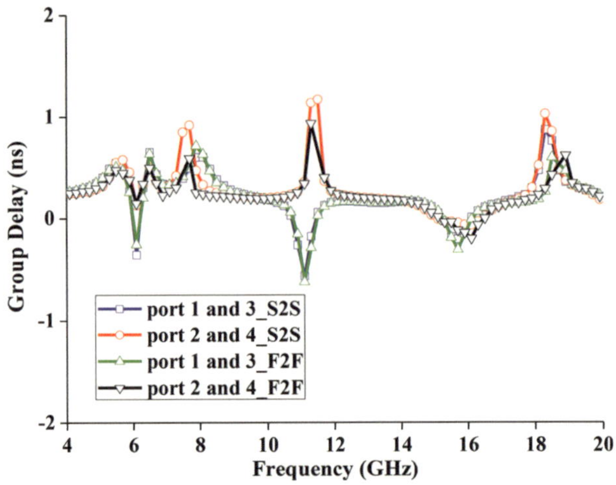

Figure 12.
Group delay versus frequency characteristics of the diversity antenna for all four configurations.

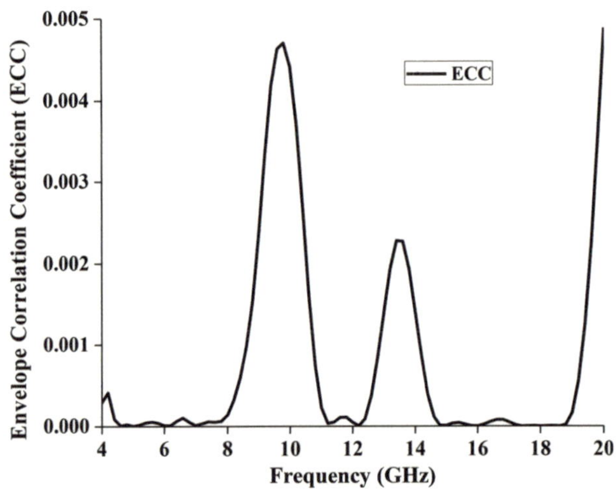

Figure 13.
Envelope correlation coefficient versus frequency characteristic.

The mean effective gain measures the antenna gain of each antenna element taking the radiation power pattern effects, the antenna total efficiency and the propagation effects into account. It is calculated by using Eq. (5) and is plotted for each antenna in **Figure 14**.

$$MEG = \int_0^{2\pi}\int_0^{\pi}\left[\frac{XPR}{1+XPR}G_\theta(\theta,\varphi)P_\theta(\theta,\varphi)+\frac{XPR}{1+XPR}G_\varphi(\theta,\varphi)P_\varphi(\theta,\varphi)\right]\sin\theta d\theta d\varphi \tag{5}$$

where XPR represents the cross-polarization ratio, $G_\theta$and $G_\varphi$are the θ- and φ-components of the antenna power gain patterns and P_θ and P_φ are the θ- and φ-components of the angular density functions of the incident power, respectively. The MEG values for each antenna element in the case of isotropic radiation, i.e. XPR = 0 dB, are presented in **Figure 14**.

Another important parameter used to identify the suitability of an antenna for diversity applications is diversity gain. It is the difference between the selection combined cumulative distribution function (CDF) and one of the other CDFs at a certain CDF level. The commonly used CDF level is 1% [33]. The DG of the diversity antenna can be calculated approximately by Eq. (6) [34]. From **Figure 15**, it is observed that the diversity gain value is almost constant in the entire band of operation.

$$DG = 10\sqrt{1-\rho_e} \tag{6}$$

In the case of a rich multipath environment, the maximum rate of transmission for reliable transmission in a communication channel is estimated by calculating capacity loss (b/s/Hz). For a MIMO antenna, a channel capacity loss of less than 0.4 b/s/Hz is acceptable [35]. It is calculated by using the correlation matrix (7) [35].

$$C_{loss} = -\log_2\left(\psi^R\right) \tag{7}$$

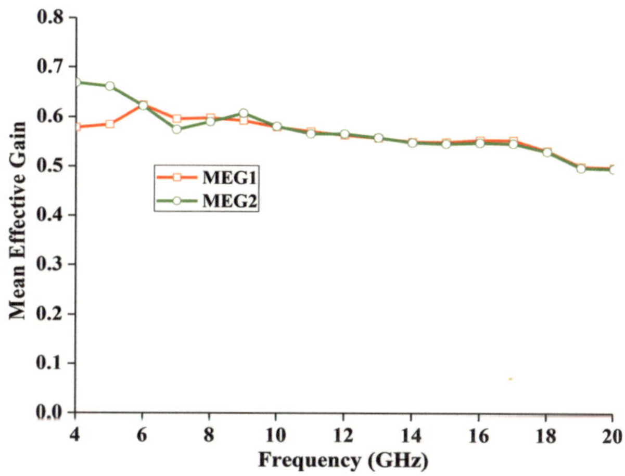

Figure 14.
Mean effective gain versus frequency characteristics of the designed antenna.

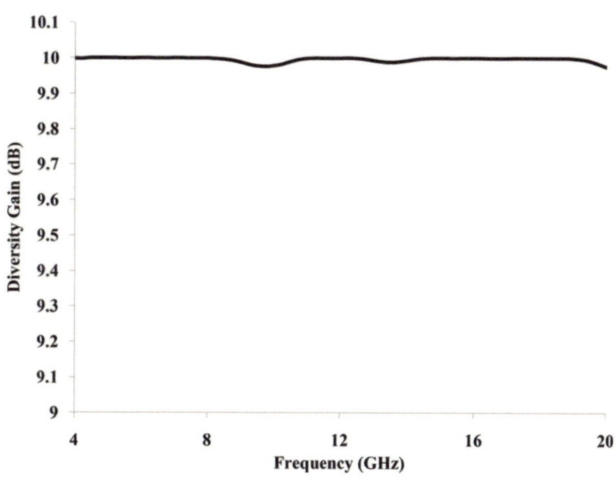

Figure 15.
Diversity gain versus frequency characteristic.

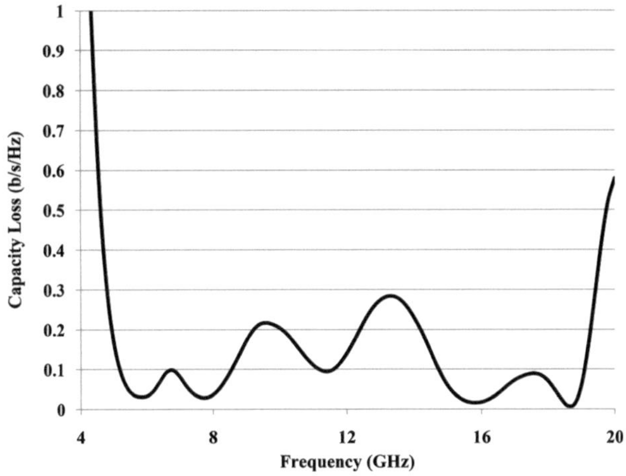

Figure 16.
Capacity loss versus frequency characteristic.

where ψ^R is the correlation matrix of the receiving antenna and is expressed mathematically as

$$\psi^R = \begin{bmatrix} \rho_{11} & \rho_{12} \\ \rho_{21} & \rho_{22} \end{bmatrix} \quad (8)$$

$$\rho_{ii} = 1 - \left(|S_{ii}|^2 + |S_{ij}|^2\right) \quad (9)$$

$$\rho_{ij} = -\left(S_{ii}^* S_{ij} + S_{ji}^* S_{ij}\right) \quad (10)$$

i, j = 1 or 2.

Figure 16 shows that the channel capacity loss changes with the variation of frequencies. It can be seen that the capacity loss values are always less than 0.3 b/s/Hz in the UWB operating range.

4. Conclusion

A compact third iteration fractal antenna with notch-loaded ground plane for UWB applications is investigated and analysed. By using orthogonal arrangement of two antenna elements, polarization diversity performance is achieved. The designed antenna has an impedance bandwidth of 4.7–19.4 GHz, isolation of more than 15 dB, a size miniaturization up to 92% over previously reported structures and good diversity performance parameters values like ECC < 0.005, almost constant DG = 10 and channel capacity loss of less than 0.3 b/s/Hz which makes it suitable for UWB polarization applications in future wireless communication systems to mitigate the multipath fading.

Author details

Sarthak Singhal[1*] and Amit Kumar Singh[2]

1 Department of Electronics and Communication Engineering, Malaviya National Institute of Technology, Jaipur, Rajasthan, India

2 Department of Electronics Engineering, Indian Institute of Technology (BHU), Varanasi, Uttar Pradesh, India

*Address all correspondence to: sarthak.ece@mnit.ac.in

References

[1] Federal Communication Commission. "First order and report: Revision of part 15 of the Commission's rules regarding UWB transmission systems," April 22, 2002

[2] Dietrich CB Jr, Dietze K, Nealy JR, Stutzman WL. Spatial, polarization, and pattern diversity for wireless handheld terminals. IEEE Transactions on Antennas and Propagation. 2001; **49**(9):1271-1281

[3] Allen B, Dohler M, Okon E, Malik WQ, Brown AK, Edwards D. UWB Antenna and Propagation for Communications, Radar and Imaging. Hoboken, NJ, USA: Wiley; 2007

[4] Wang A, Zhenghe F, Luk K-M. Pattern and polarization diversity antenna with high isolation for portable wireless devices. IEEE Antennas and Wireless Propagation Letters. 2009;**8**: 209-211

[5] Naraghi-Pour M, Ikuma T. "Diversity techniques for spectrum sensing in fading environments," presented at the MILCOM; 2008

[6] Gkonis P, Tsoulos G, Kaklamani D. Dual code Tx diversity with antenna selection for spatial multiplexing in MIMO-WCDMA networks. IEEE Communications Letters. 2009;**13**(8): 570-572

[7] Boyle K. Radiation patterns and correlation of closely spaced linear antennas. IEEE Transactions on Antennas and Propagation. 2002;**50**(8): 1162-1165

[8] Wei K, Zhang Z, Chen W, Feng Z. A novel hybrid-fed patch antenna with pattern diversity. IEEE Antennas and Wireless Propagation Letters. 2010;**9**: 562-565

[9] Yang SLS, Luk K-M, Lai H-W, Kishk A-A, Lee K-F. A dual-polarized antenna with pattern diversity. IEEE Antennas and Propagation Magazine. Dec. 2008; **50**(6):71-79

[10] Huang Y, Nehorai A, Friedman G. Mutual coupling of two collocated orthogonally oriented circular thin-wire loops. IEEE Transactions on Antennas and Propagation. 2003;**51**(6): 1307-1314

[11] Gallo M, Daviu EA, Bataller MF, Bozzetti M, Pardo JMMG, Llacer LJ. A broadband pattern diversity annular slot antenna. IEEE Transactions on Antennas and Propagation. 2012;**60**(3): 1596-1600

[12] Bao XL, Ammann MJ. Wideband dual-frequency dual-polarized dipole-like antenna. IEEE Antennas and Wireless Propagation Letters. 2011;**10**: 831-834

[13] Adamiuk G, Zwick T, Wiesbeck W. Compact, dual-polarized UWB-antenna, embedded in a dielectric. IEEE Transactions on Antennas and Propagation. 2010;**58**(2):279-286

[14] Yoon HK, Yoon YJ, Kim H, Lee CH. Flexible ultra-wideband polarisation diversity antenna with band-notch function. IET Microwaves, Antennas and Propagation. 2011;**5**(12):1463-1470

[15] Chacko BP, Augustin G, Denidni TA. Uniplanar slot antenna for ultrawideband polarization diversity applications. IEEE Antennas and Wireless Propagation Letters. 2013;**12**: 88-91

[16] Koohestani M, Moreira AA, Skrivervik AK. A novel compact CPW-fed polarization diversity ultrawideband antenna. IEEE Antennas and Wireless Propagation Letters. 2014; **13**:563-566

[17] Femina BS, Mishra SK. Compact WLAN band-notched printed ultrawideband MIMO antenna with polarization diversity. Progress in Electromagnetics Research. 2016;**61**: 149-159

[18] Chacko BP, Augustin G, Denidni TA. Uniplanar polarisation diversity antenna for ultrawideband systems. IET Microwaves, Antennas and Propagation. 2013;**7**(10):851-857

[19] Khan MS, Capobianco AD, Naqvi A, Ijaz B, Asif S, Braaten BD. Planar, compact ultra-wideband polarisation diversity antenna array. IET Microwaves, Antennas and Propagation. 2015;**9**(15):1761-1768

[20] Daviu EA, Gallo M, Clemente BB, Bataller MF. Ultrawideband slot ring antenna for diversity application. Electronics Letters. 2010;**46**(7):478-480

[21] Srivastava K, Kumar A, Kanaujia BK, Dwari S, Kumar S. A CPW-fed UWB MIMO antenna with integrated GSM band and dual band notches. International Journal of RF and Microwave Computer-Aided Engineering. 2018;**29**(1):1-10

[22] Biswal SP, Das S. A compact dual port UWB-MIMO/diversity antenna for indoor application. International Journal of Microwave and Wireless Technologies. 2017;**10**(03):360-367

[23] Biswal SP, Das S. A compact printed ultra-wideband multiple-input multiple-output prototype with band-notch ability for WiMAX, LTEband43, and WLAN systems. International Journal of RF and Microwave Computer-Aided Engineering. 2019;**6**:1-11

[24] Ibrahim AA, Machac J, Shubair RM. Compact UWB MIMO antenna with pattern diversity and band rejection characteristics. Microwave and Optical Technology Letters. 2017;**59**(6): 1460-1464

[25] Iqbal A, Saraereh OA, Ahmad AW, Bashir S. Mutual coupling reduction using F-shaped stubs in UWB-MIMO antenna. IEEE Access. 2018;**6**:2755-2759

[26] Zhao X, Yeo SP, Ong LC. Planar UWB MIMO antenna with pattern diversity and isolation improvement for Mobile platform based on the theory of characteristic modes. IEEE Transactions on Antennas and Propagation. 2018; **66**(1):420-425

[27] Yadav D, Abegaonkar MP, Koul SK, Tiwari VN, Bhatnagar D. Two element band-notched Uwb Mimo antenna with high and uniform isolation. Progress in Electromagnetics Research. 2018;**63**: 119-129

[28] Sipal D, Abegaonkar MP, Koul SK. Compact dual band-notched UWB MIMO antenna for USB dongle application with pattern diversity characteristics. Progress in Electromagnetics Research. 2018;**87**: 87-96

[29] Singhal S, Goel T, Singh AK. Inner tapered tree shaped fractal antenna for UWB applications. Microwave and Optical Technology Letters. 2015;**57**(3): 559-567

[30] HFSS, "High Frequency Structure Simulator ver. 11, Finite Element Package."Ansoft Corporation. 2009. Available from: http://www.ansoft.com

[31] CST Microwave Studio Suite 2011. CST Inc.; 2007

[32] Blanch S, Romeu J, Corbella I. Exact representation of antenna system diversity performance from input parameter description. Electronics Letters. 2003;**39**(9):705-707

[33] Kildal PS, Rosengran K. Correlation and capacity of MIMO systems and coupling, radiation efficiency, diversity gain of their antennas: Simulations and measurements in a reverberation

chamber. IEEE Communications Magazine. 2004;**42**(12):102-112

[34] Gao Y, Chen XD, Ying ZN. Design and performance investigation of a dual-element PIFA array at 2.5 GHz for MIMO terminal. IEEE Transactions on Antennas and Propagation. 2007;**55**(12): 3433-3441

[35] Tripathi S, Mohan A, Yadav S. A compact Koch fractal UWB MIMO antenna with WLAN band-rejection. IEEE Antennas and Wireless Propagation Letters. 2015;**14**:1565-1568

Chapter 4

Antennas for UWB Applications

Symeon Nikolaou and Abdul Quddious

Abstract

"Antennas for UWB Applications" chapter deals with an overview of ultra-wideband (UWB) antennas used for different applications. Some fundamental and widely used radiators, such as fat monopole, microstrip-fed and coplanar waveguide (CPW)-fed slot antennas, and tapered end-fire antennas are presented. Selected antenna designs are presented in relation to the UWB applications and their dictating radiation and operation principles. The demonstrated UWB antennas include antennas for handheld devices used for personal area network (PAN) communications; antennas for localization and positioning; UWB antennas for radio-frequency identifications (RFIDs); radar antennas for through-wall imaging, for ground-penetrating radar (GPR), and for breast tumor detection; and more generally, UWB antennas used for sensing. For some of the aforementioned applications, UWB antennas with special characteristics are needed, and they are presented and associated with the relevant applications. These include reconfigurable UWB antennas, metamaterial-loaded UWB antennas, and conformal UWB antennas. The usefulness of these special characteristics in comparison with the claimed advantages is critically evaluated. For the UWB applications presented in the chapter, one type or UWB antenna is recommended.

Keywords: UWB antennas, reconfigurable UWB antennas, UWB-RFID, conformal UWB antennas, microwave imaging, breast tumor detection

1. Introduction

There is a wide range of fundamentally different applications that use UWB antennas, and as a general approach, different customized antennas are needed depending on the desired radiation characteristics. The most common applications are data communication, localization and identification, and radar and sensing applications.

UWB communications have inherently very wide bandwidth in which, based on Shannon's theorem, these systems can support high data rates. Therefore, UWB transceivers are used for the transmission of high data rate, wireless communications, which are often used for PAN communications. One popular commercial application is the wireless USB (WUSB) which is designed to achieve 480 Mbit/s at distances up to 3 m and 110 Mbit/s at up to 10 m. PAN communications involve mobile handheld devices, in which case the used antennas should preferably have omnidirectional patterns with compact size and planar designs [1, 2]. Considering that UWB spectrum is shared with other technologies and standards such as the 3.6 GHz IEEE 802.11y wireless local area networks (WLAN) (3.6575–3.69 GHz), 4.9 GHz public safety WLAN (4.94–4.99 GHz), and 5 GHz IEEE 802.11a/h/j/n WLAN (5.15–5.35, 5.25–5.35, 5.47–5.725, 5.725–5.825 GHz), all operating within

IntechOpen

the Federal Communication Commission (FCC) designated UWB band, of 3.1–10.6 GHz, the design of reconfigurable notch-band antennas has attracted a lot of attention [3–5] since they can potentially filter out the unwanted interferer.

UWB technology is used for positioning and location tracking. In the general principle, a UWB interrogator transmits a signal which is reflected by UWB tags which are identified, and depending on the number of interrogators and the utilized software, the position of the specific tag can be defined with relatively high accuracy. Currently there are companies such as UWINLOC which offer integrated solutions for smart and efficient asset management through real-time location systems (RTLS) by combining UWB technology with Internet of things (IoT) principles. Depending whether the UWB antenna is intended for the interrogator [6] or the UWB-RFID tag [7], the radiation characteristics and the size constraints may vary significantly. While the interrogator can combine multiple elements in arrays with beam-forming capabilities, the RFID tags need to be compact, lightweight, omnidirectional, and mostly, low-cost. In order to meet this last requirement, chipless UWB RFIDs [8] are used since they can be easily and massively manufactured on demand using additive fabrication technology.

"Radar devices" involve a wide variety of highly specialized applications for which UWB technology and UWB antennas are widely used even if many of the preferred UWB antennas radiate on different frequencies than the FCC designated 3.1–10.6 GHz band. Ground-penetrating radars (GPR) is one such application for which radars are used either to detect objects buried in the ground [9], to estimate soil characteristics (i.e., moisture) [10], or even to detect living beings trapped in ruins, after a physical disaster such as an earthquake or a hurricane. For this latter case, a more sophisticated radar—through-wall imaging radar—can also be used. Through-wall imaging systems are usually limited for use from law enforcement units, to monitor the position and movements of potentially dangerous targets [11], or even as airport security measure, in order to identify concealed weapons [12]. Microwave imaging [13] which is a more general category, under which the two aforementioned applications can be classified, includes the medical microwave imaging category which has attracted a lot of attention in the recent years. Medical microwave imaging is widely used for breast tumor detection [14, 15], and after several research efforts that investigated conformal UWB arrays [16, 17] and image reconstruction algorithms, several university spin-offs and other private companies proceeded with the implementation of commercial devices that were cleared for clinical studies with human subjects [18, 19]. These devices include both large-size devices facilitated at hospitals [20] and lightweight wearable devices for individual use at home [21].

This chapter presents an indicative list of UWB antennas which can be used for data communications, RFIDs used for identification and localization, and UWB antennas used for sensing and radar applications.

2. UWB antennas for wireless communications

"UWB communication systems" include communications which serve a variety of purposes. In a more loose definition, ultrawideband communication system is a system that requires more than 500 MHz bandwidth. Consequently, applications can be found in several subsections of the electromagnetic spectrum, starting from the UHF band all the way up to mm wave and sub-mm wave frequencies. This section focuses on UWB antennas for communications within the FCC designated UWB frequency range, 3.1–10.6 GHz. When UWB communications started to concentrate increased researchers' attention, (First International Conference on UWB-ICUWB, in 2001, is considered a milestone), in the early 2000, the used antennas were rather large in size. Some of the earlier UWB antenna versions included self-complementary antennas which are frequency independent (spiral antenna) or wideband dipoles [22]. For the use of UWB technology for mobile handheld devices, the large antenna size was becoming a limiting factor. Consequently, size reduction became one of the primary objectives of the UWB antenna designers.

In the next years, self-complementary antennas and end-fire antennas were replaced with multiple resonator antennas. The design approach is to create several resonances on the S_{11} which are distributed along the 3.1–10.6 GHz frequency range. As a result, the reflection coefficient is "pushed" lower than −10 dB conventional matching threshold which is widely used as the "good matching" criterion. The existence of radiating segments of specific size causes the creation of the resonances in the reflection coefficient, and the size variation of the physical size of the resonators results in the corresponding resonance shift in frequency. **Figure 1** illustrates a series of multiple resonator antennas which progressively limit their overall (board) size. The antennas are CPW-fed, and the central (signal) conductor is linearly tapered to improve the matching. Details about the design and performance of the depicted antennas can be found in [2]. It has been shown that the transition from the conventional transmission line to the radiator and the ground size both affect the overall matching and consequently the radiation efficiency of the UWB antenna and should be considered by antenna designers as inherently integrated parts of the radiator itself. **Figure 1a** demonstrates an uneven U-shaped stub, radiating inside an elliptical slot. Under the observation of the limited effect of the slot on the matching, the slot was removed, and an additional linear segment was added on

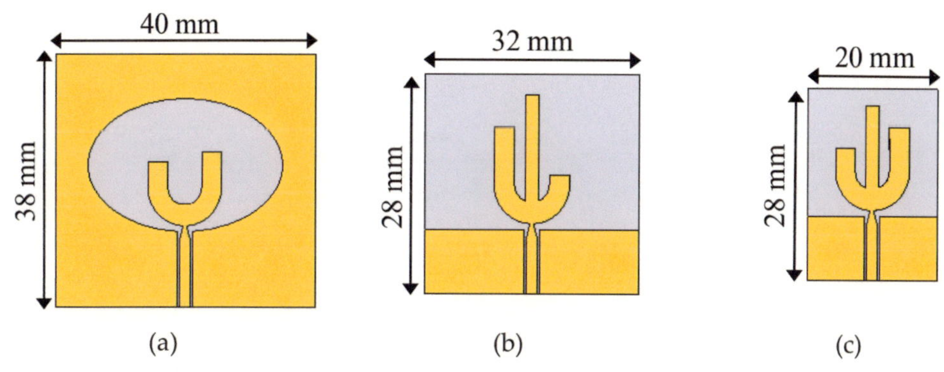

Figure 1.
CPW-fed (a) elliptical slot, (b) cactus and (c) compact-cactus UWB antenna.

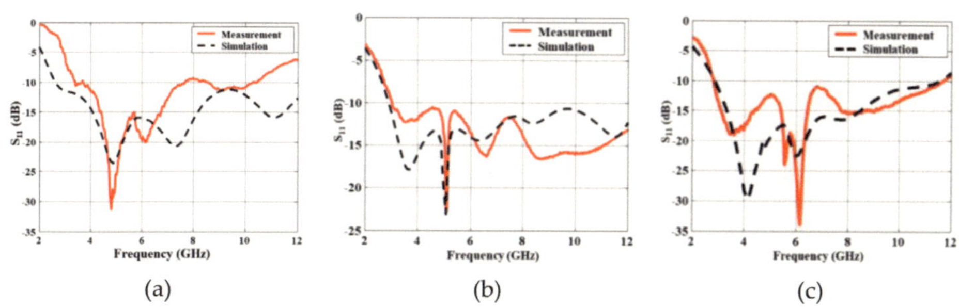

Figure 2.
Simulated and measured S_{11} for (a) CPW-fed slot antenna, (b) cactus antenna, and (c) compact-cactus antenna [22].

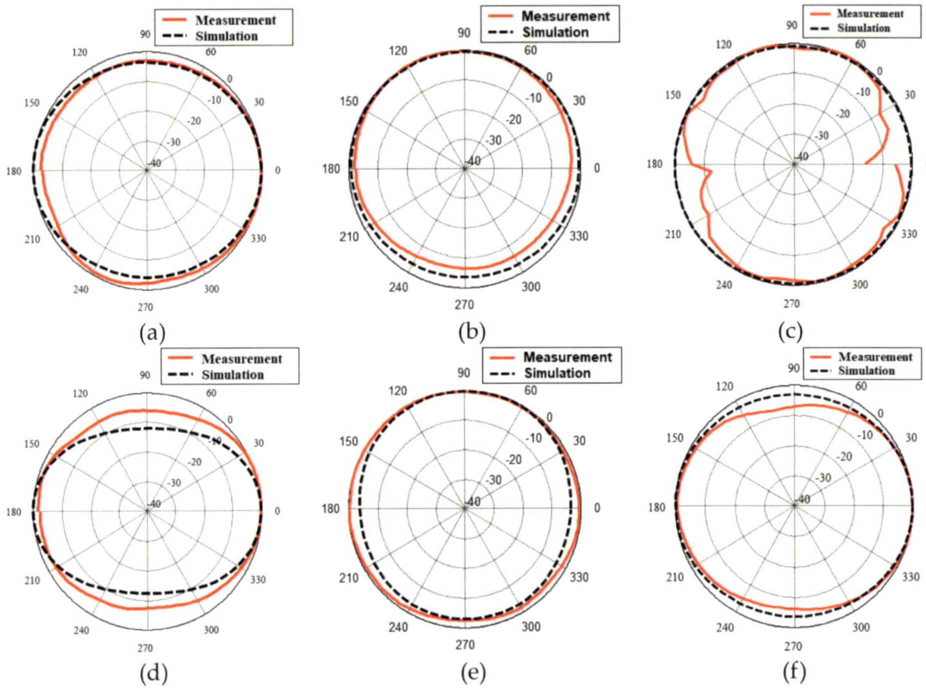

Figure 3.
Simulated and measured H-plane radiation patterns of (a) CPW-fed slot antenna, (b) cactus antenna, and (c) compact-cactus antenna at 5 GHz and (d) CPW-fed slot antenna, (e) cactus antenna, and (f) compact-cactus antenna at 9 GHz [22].

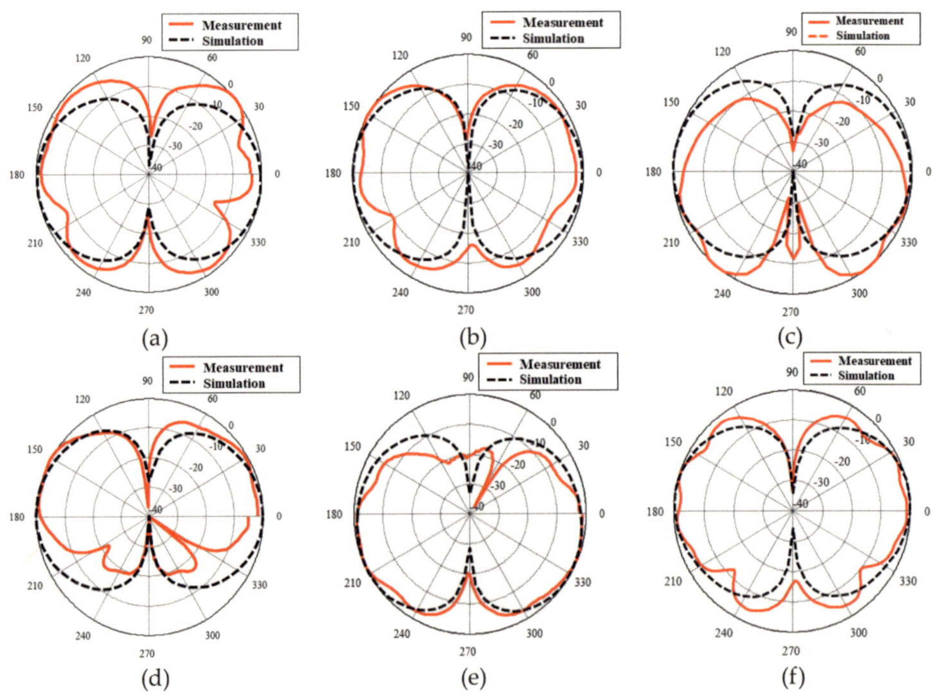

Figure 4.
Simulated and measured E-plane radiation patterns of (a) CPW-fed slot antenna, (b) cactus antenna, and (c) compact-cactus antenna, at 5 GHz, and (d) CPW-fed slot antenna, (e) cactus antenna, and (f) compact-cactus antenna, at 9 GHz [22].

the U-shaped stub, converting it into a "cactus-shaped" radiator consisting of three linear segments (**Figure 1b**). The last iteration presents a variation of the cactus-shaped radiator, the "compact cactus," with smaller RF-ground patches (**Figure 1c**) and further decreased overall size. **Figure 2** presents the reflection coefficient plots where the resonances are distributed along the entire UWB frequency band.

Although the elliptical slot does not affect the matching significantly, it has more profound impact on the resulted radiation patterns. The removal of the slot modifies the radiation patterns to rather omnidirectional patterns along the H-plane (**Figure 3**), while the E-plane (**Figure 4**) patterns are characterized by a null in the direction of the feeding line. Several other published research works demonstrated the attractive features and the effectiveness of the so-called fat monopole or the elliptical disc and its variations, as UWB radiator, and it became one of the most widely used UWB antenna types, for mobile handheld devices.

3. Reconfigurable UWB antennas for wireless communications

UWB systems share the same spectrum with several other narrowband wireless systems which use sub-bands inside the 3.1–10.6 GHz range. The FCC mask limits the UWB EIRP to −41.3 dBm/MHz, which means that UWB signals are rather weak, to degrade the performance of the narrowband systems, significantly. UWB signals are considered white noise for narrowband systems. On the other hand, UWB systems suffer from the strong interfering signals which are used from the narrowband systems. In order to reduce the received noise level and improve the associated SNR, UWB antennas are designed with notch bands, which effectively filter out the received signals at the frequencies used from a competing narrowband system. Ideally these notch bands should be reconfigurable, in other words to appear when an interfering signal is detected and to disappear when no such signal is detected. Generally, notch bands are caused from added resonators which are placed on the radiator, or the feeding line, or even the RF ground patches. **Figure 5** presents a microstrip-fed monopole with three pairs of added resonators which cause three notch bands on the reflection coefficient [23]. Each pair of resonators is designed to cause a notch band at a desired frequency. Specifically, the two stepped λ/4 open stubs cause the notch at the WiMAX (3.5 GHz) band; the two capacitively loaded loops (CLLs) on the ground patch cause the notch at the WLAN (5.8 GHz) band, and the pair of linear λ/2 segments, printed on the back side of the radiator, causes the notch at 8.87 GHz. The use of three different pairs of resonators allows the control of the notch bands independently. The effect of the resonators can be made reconfigurable if suitable switches are used at the right place. **Figure 5b** shows that if the λ/4 open stubs are disconnected (switch in off state), the frequency notch disappears (blue solid line). In **Figure 5c** the reflection coefficient of the UWB antenna is presented when the switch on the CLL is in either on or off state. With the switch in off state, the WLAN notch exists and filters out the undesired high-power signal, while when the switch is in on state, the effect of the CLL is canceled, and as a result the UWB antenna radiates efficiently at the WLAN band. Finally, when the switch separating the λ/2 linear segment into two unequal parts (**Figure 5d**) is set to off state, it causes the cancelation of the notch at 8.8 GHz.

Depending on the geometry of the resonator, a variety of electronic switches can be implemented. **Figure 6** presents three implemented UWB antennas [24–25], with reconfigurable notch bands at the WLAN band (5.8 GHz) which use a single resonator instead of a pair of resonators and, consequently, only one switch to implement the notch reconfigurability feature. A microstrip-fed monopole with a J-shaped stub inside a rectangular slot is presented in **Figure 6a**, where the J-shaped

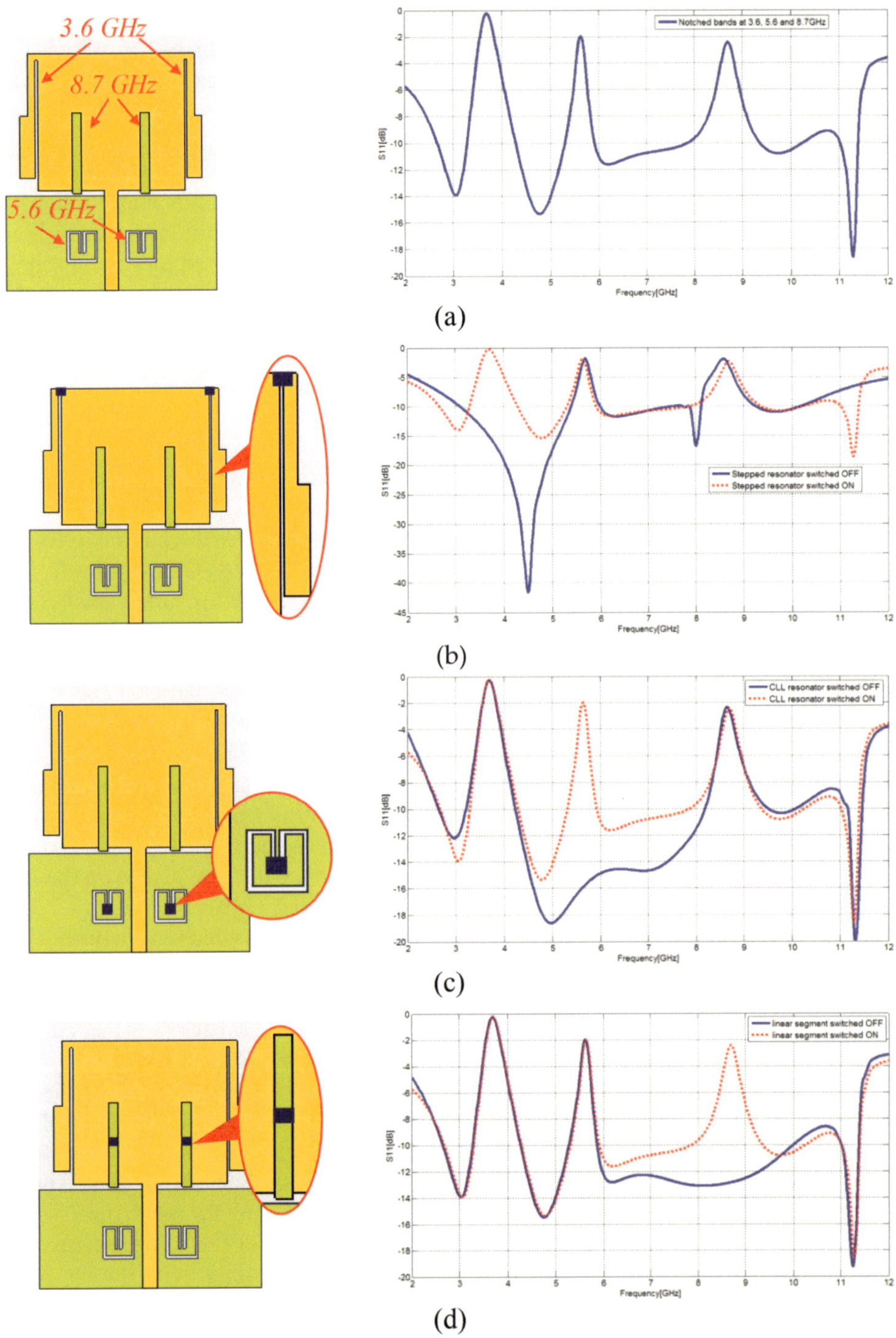

Figure 5.
Simulated S_{11} microstrip-fed UWB (a) monopole with three pairs of resonators (b), stepped λ/4 open stubs (c), CLL resonators, and λ/2 parasitic linear segments (d).

stub is connected and disconnected to the radiator, using a PIN diode switch. The J-shaped stub causes the notch, but when the diode is set to off state, the stub is disconnected, and the frequency notch is suppressed. In a similar design, the dynamically reconfigurable UWB antenna presented in **Figure 6b** uses a low-power field-effect transistor as switch (FET switch) to connect and disconnect the linear

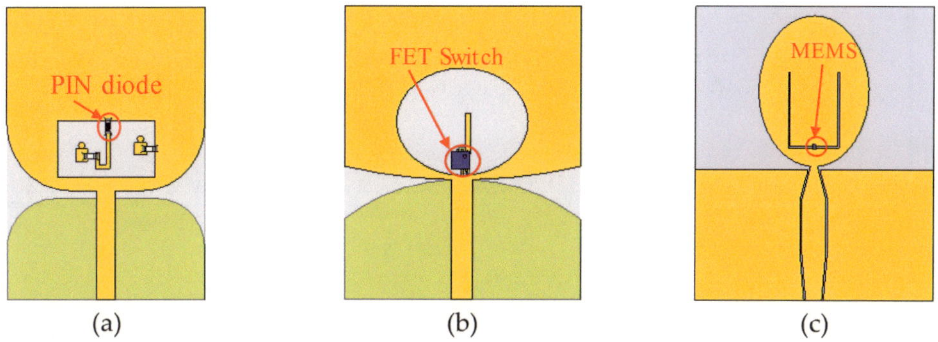

Figure 6.
Reconfigurable UWB antennas with (a) PIN diode, (b) FET switch, and (c) MEMS.

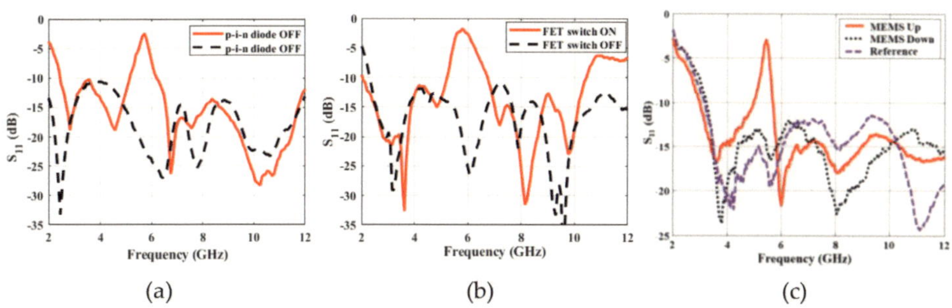

Figure 7.
Measured S_{11} of three different reconfigurable UWB antennas with (a) PIN diode, (b) FET switch, and (c) MEMS.

segment inside the elliptical slot. The FET switch can be actuated without battery, using solely the ambient RF power which is collected from a 5.8 GHz rectenna. Finally, **Figure 6c** presents a CPW-fed elliptical slot which uses a U-shaped slot with a microelectromechanical system (MEMS) switch to implement the frequency notch reconfigurability. The MEMS switch is actuated without bias lines, while for the application of the differential voltage for the reconfigurable UWB antennas presented in **Figure 6a** and **b**, bias lines with RF choke inductors are necessary. The MEMS switch was fabricated in-house [25], while the PIN diode and the FET switch are off-the-shelf components. (**Figure 7**) presents the measured S_{11} of three different reconfigurable UWB antennas that show successful implementation of a reconfigurable notch band at 5.8 GHz.

4. UWB antenna for chipless RFIDs

UWB monopoles are also used for the implementation of chipless UWB RFIDs [8]. Chipless tags are either backscattering-based or retransmission-based. Usually, UWB RFIDs are retransmission-based, and on-off keying (OOK) is performed by the presence or absence of a series of resonators which are coupled with the transmission line. Alternatively, the resonators may be perturbed and thus detuned by either short circuiting or open circuiting them. Backscattering occurs when a single antenna with high Q is used, while retransmission requires a second antenna which is preferably orthogonally polarized compared to the receiver, and it transmits a modulated signal after the unmodulated signal is received from the receiver antenna. The OOK modulation in chipless tags is performed in the frequency domain, and an

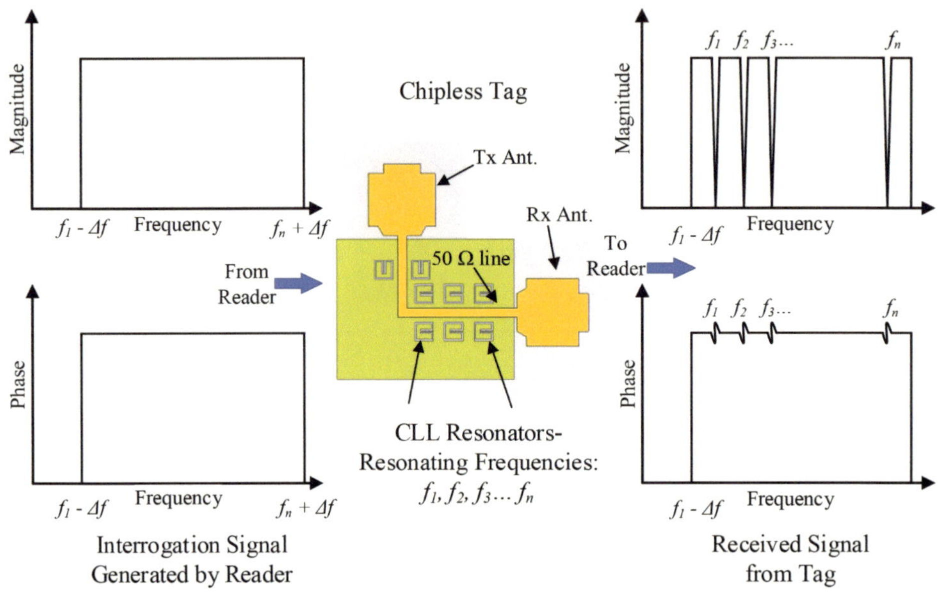

Figure 8.
Operation principle and schematic of chipless UWB RFIDs.

indicative signature of a retransmitted signal can be seen in **Figure 8**. The chipless UWB RFID schematic that is presented in **Figure 8** consists of a pair of UWB monopoles which are connected with a common feeding line. The feeding line is loaded with eight resonators forming an 8-bit word. Each resonator which has a slightly different size causes a frequency notch at a different frequency, and at the same time, it causes a phase discontinuity. With the use of a UWB interrogator, a wideband signal is sent, and the tag receives the signal, and it retransmits it back to the reader. The combination of resonators causes a unique electromagnetic signature, which identifies the tag. There is a variety of resonators that can be used, such as slits, CLLs, or slot ring resonators (SRRs). Batteryless, chipless, and entirely passive UWB RFIDs can be inkjet-printed on paper substrates and be massively and rapidly produced for disposable RFID tags, such as the baggage paper-based tags which are used to identify the checked-in luggage. A pair of microstrip-fed UWB monopole antennas is well-suited for the implementation of chipless UWB RFIDs.

5. UWB antennas for radar applications

In many types of radars such as GPR, through-wall imaging radars, or even breast imaging systems, UWB signals are required. For high-gain and high-power radars, end-fire UWB antennas are preferred because they perform with unidirectional, high gain which is necessary for enhanced radar range. Linearly tapered antennas (LTA) like Vivaldi antennas [26, 27] are widely used, and the more complex double exponentially tapered slot antennas (DETSA) can also be used. **Figure 9** presents a high-gain DETSA antenna fabricated on flexible liquid crystal polymer (LCP) substrate [28]. The implemented gain varies between 7 and 12 dBi. The use of flexible substrate such as LCP allows the adoption of the longitudinal antenna along a non-planar surface as can be seen in **Figure 9b**. End-fire DETSA has been shown to redirect the maximum gain direction along the tangent of the ending point; therefore, it can be used to implement high-gain directive radar front ends where two or even four DETSA antennas can be combined to form a confocal high-gain

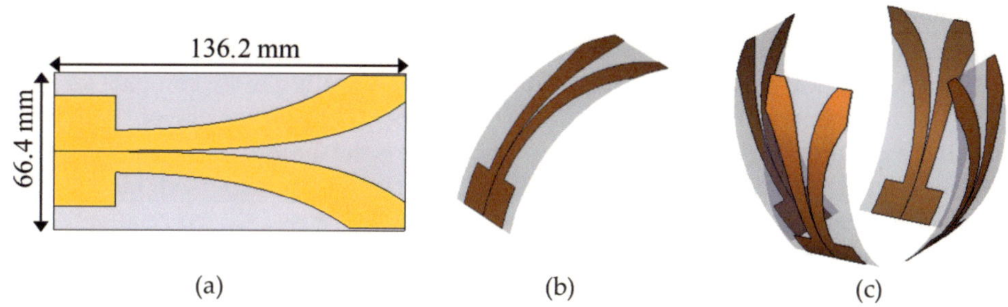

Figure 9.
(a) Planar DETSA, (b) conformal DETSA, (c) DETSA antennas confocal array.

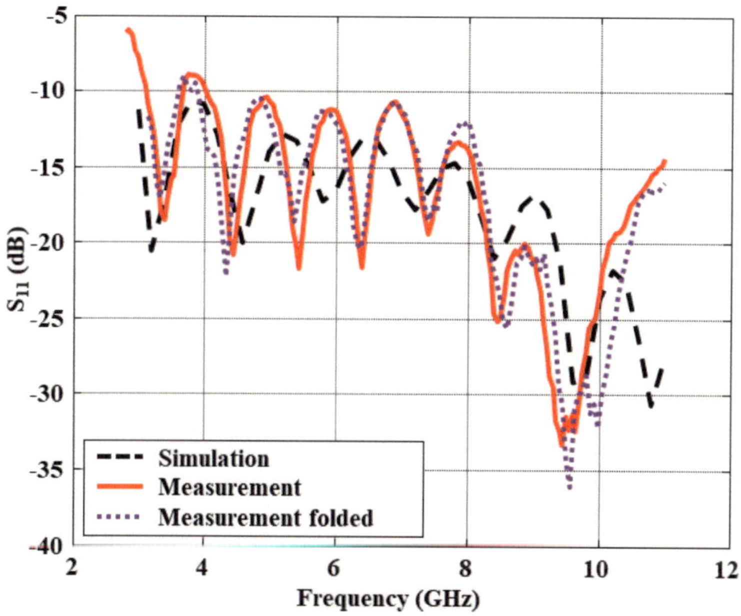

Figure 10.
Simulated and measured S_{11} for planar and folded DETSA [22].

array as can be seen in **Figure 9c**. For GPR or through-wall imaging radars, a directive confocal array is placed about the ground target or on the wall. These radars are mono-static, and the same antenna system operates both as the transmitter and the receiver. It is desired to have high penetration capability which means that high-gain antennas and high-gain amplifiers are used to detect reflected signals propagating through high-loss media. End-fire high-gain antennas such as the Vivaldi antennas and their variations are usually used for these applications. In (**Figure 10**) it can be seen that good agreement is achieved between simulated and measured S11 and there is no difference between planar and folded DETSA return loss.

6. Conformal UWB antennas

In several occasions UWB antennas need to be integrated or mounted on non-planar surfaces, and consequently the use of conformal UWB antennas has been investigated by several researchers. In [22, 29] three distinct UWB antennas are used in order to test their matching and radiation characteristics when the antennas are mounted on cylindrical surfaces of different radii. The first is a polygonal monopole antenna which will be referred to with the acronym PM for brevity. The second is a CPW-fed, elliptical slot antenna or ES for brevity, and finally a more compact size elliptical monopole or EM as it will be referred as, for brevity. Since one of the most popular solutions for a variety of UWB applications is the fat monopole, two such prototypes (PM and EM), significantly different in size, are investigated. Liquid crystal polymer (LCP) is used as fabrication substrate because LCP has relatively low dielectric constant ($\varepsilon_r = 3$) and low loss ($\tan\delta = 0.002$). For the two monopole antennas, the PM and the EM, the commercially available substrate, with thickness of 100 μm, is used. The fabricated prototypes are tested in planar shape and also when they are mounted on two different Styrofoam cylinders with radius either 25 or 15 mm. The used prototype antennas in planar shapes are presented in **Figure 11**.

The prototypes are measured in planar shape (R = inf), and then the measurements are repeated with the antennas mounted on custom-made cylinders with (a) radii 25 mm (R = 25 mm) and (b) 15 mm (R = 15 mm). The used cylinders were custom-made at a machine shop, and Styrofoam ($\varepsilon_r = 1.03$) material was used to resemble free space radiation conditions. When an antenna is kept in planar shape, it may be assumed that it is mounted on a cylinder with infinite radius, and this is the description (R = inf) used in the legends for the following figures. The measured S_{11} results for the three antennas are presented in **Figure 12**. In every one of the three cases presented in **Figure 11**, it can be observed that the measured reflection coefficient for the three different radius values is almost identical. Radiation pattern measurements can be also seen in **Figures 13–15**. Considering the high accuracy and the pattern measurements, in almost every compared pair of

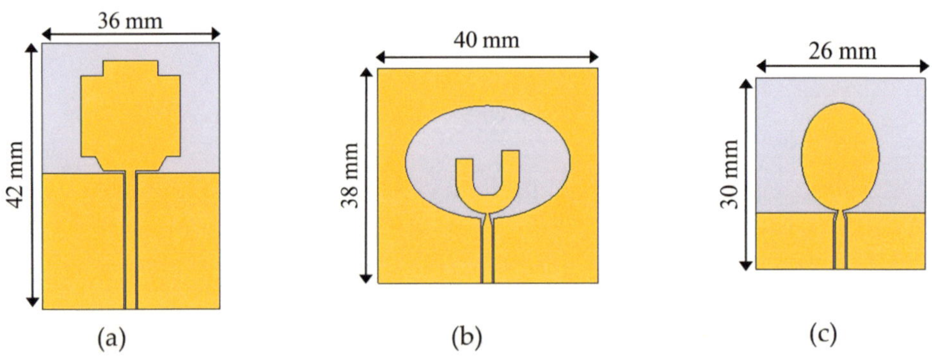

Figure 11.
Polygonal monopole (PM), elliptical slot (ES), and elliptical monopole (EM) UWB antennas in planar form.

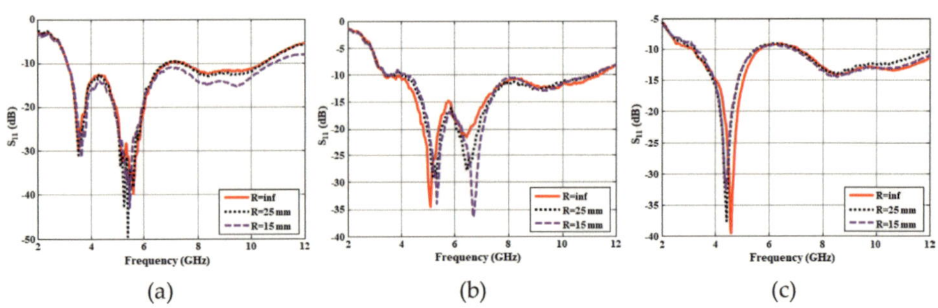

Figure 12.
Measured S_{11} under three different bending conditions for (a) polygonal monopole (PM), (b) elliptical slot (ES), and (c) elliptical monopole (EM) [22].

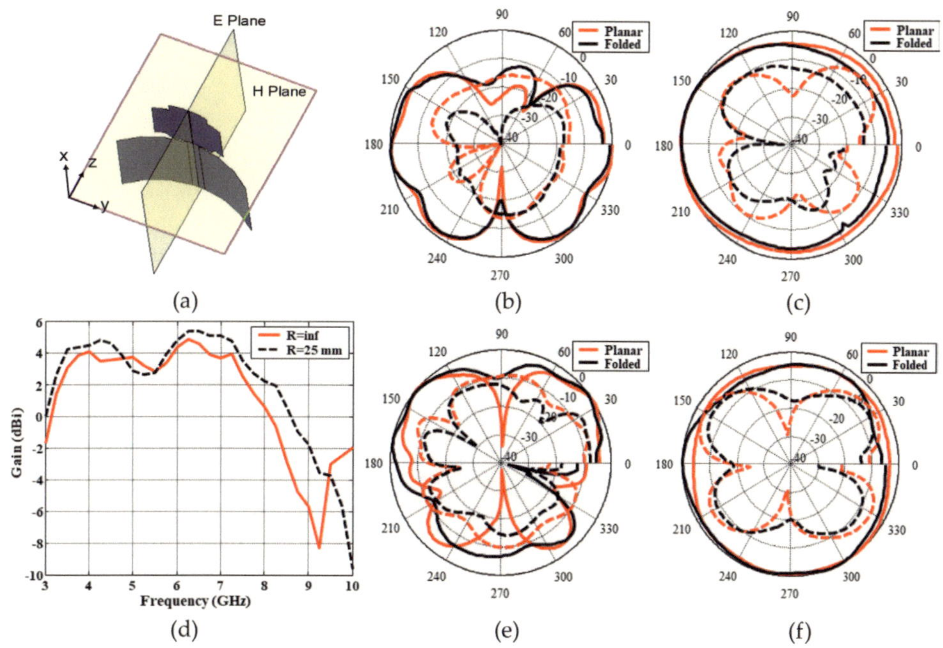

Figure 13.
Measured radiation patterns and gain for the polygonal monopole (PM) in planar form and when it is folded around a Styrofoam cylinder with a 25 mm radius. (a) Conformal PM, (b) E-plane at 5 GHz, (c) H-plane at 5 GHz, (d) measured gain (dBi) vs. frequency, (e) E-plane at 9 GHz, and (f) H-plane at 9 GHz [22].

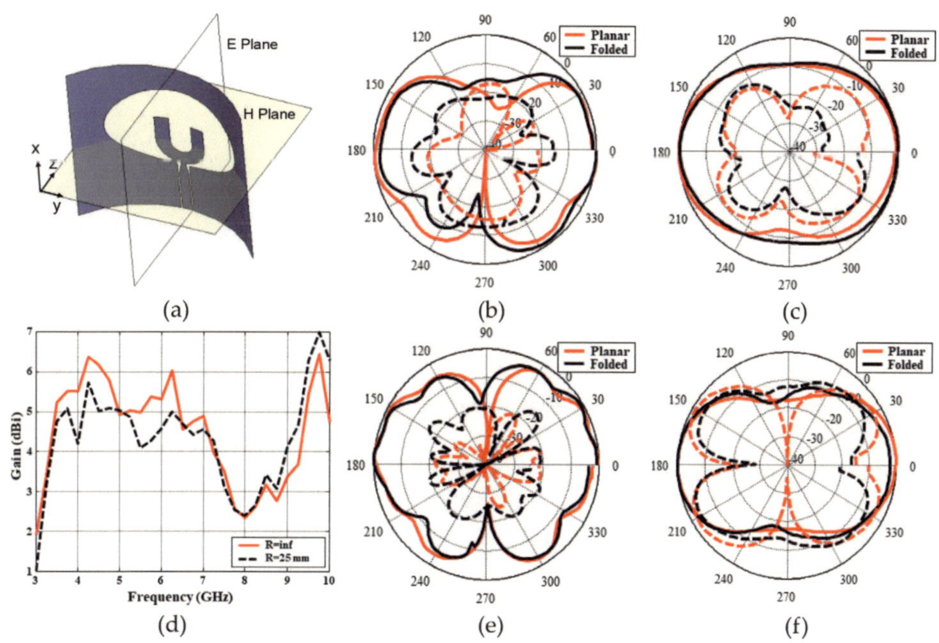

Figure 14.
Measured radiation patterns and gain for the elliptical monopole (EM) in planar form and when it is folded around a Styrofoam cylinder with a 25 mm radius. (a) Conformal EM radiator, (b) E-plane at 5 GHz, (c) H-plane at 5 GHz, (d) measured gain (dBi) vs. frequency, (e) E-plane at 9 GHz, and (f) H-plane at 9 GHz [22].

radiation patterns, the agreement between the patterns deduced from the planar and folded antennas is noteworthy. Both the shape of the patterns and the maximum directivity remain mostly steady, verifying the good agreement between the

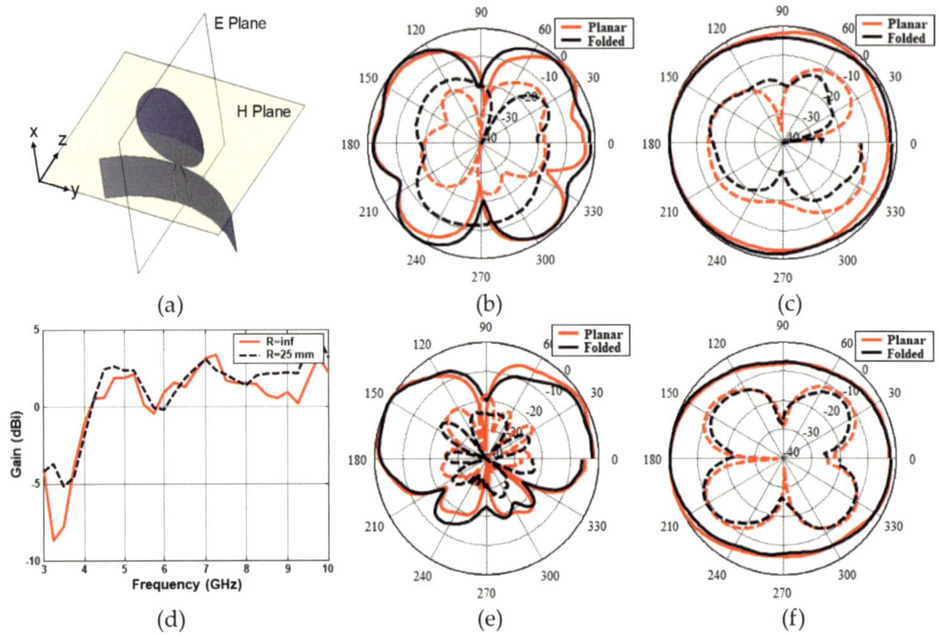

Figure 15.
Measured radiation patterns and gain for the elliptical monopole (EM) in planar form and when it is folded around a Styrofoam cylinder with a 25 mm radius. (a) EM radiator, (b) E-plane at 5 GHz, (c) H-plane at 5 GHz, (d) measured gain (dBi) vs. frequency, (e) E-plane at 9 GHz, and (f) H plane at 9 GHz [22].

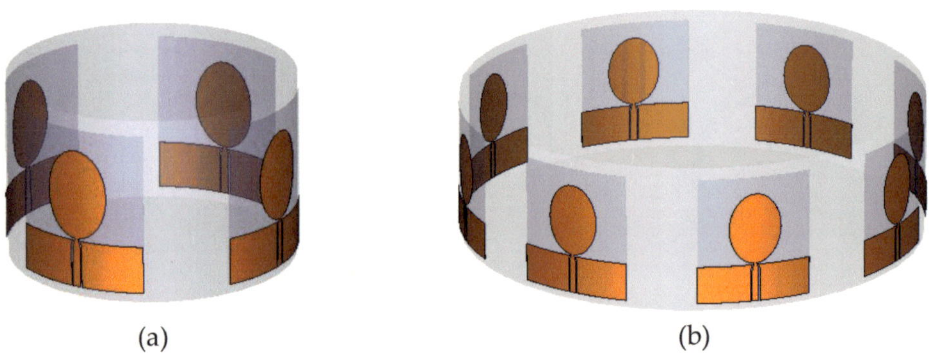

Figure 16.
Conformal cylindrical array of (a) four EM elements and (b) eights EM elements.

planar and folded antennas' radiation behavior. In conclusion and considering the testing results for all three prototypes, it can be claimed that the radiation patterns of omnidirectional UWB antennas are not significantly affected by conforming the antennas around a cylinder when the axis of the cylinder is parallel to the feeding line direction.

The use of conformal UWB antennas allows their direct use for the implementation of conformal cylindrical arrays. Such arrays are used for microwave imaging systems that are used for high-accuracy breast tumor detection devices that can be found in hospitals or even for wearable lightweight low-cost devices. **Figure 16** presents a schematic for the design of two cylindrical arrays that consist of four and eight conformal elliptical monopole UWB radiators, respectively. The radiated power is focused toward the axis of the cylinder, and the involved elements can radiate simultaneously, one at a time, or in various combinations implementing multistatic radars with different focus characteristics.

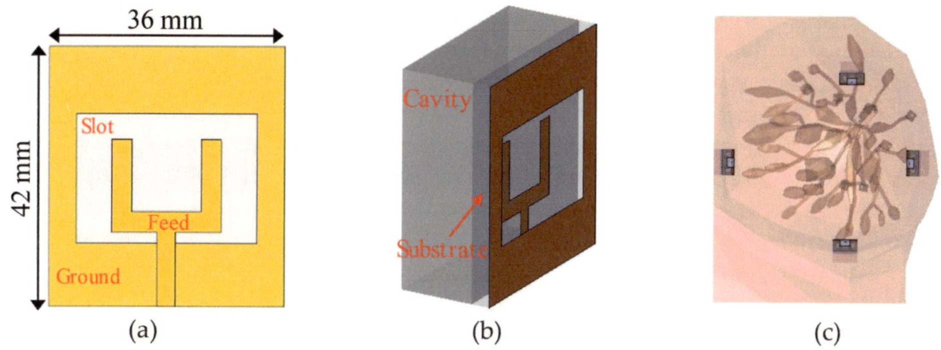

Figure 17.
(a) Cavity-backed slot UWB antenna, (b) prospective view, and (c) multiple elements in close proximity with a realistic breast phantom.

7. Microwave imaging radar antenna element

Recently microwave imaging was considered as an alternative promising technology for medical imaging especially since the required cost is much lower than the most prominent medical imaging methods such as computational tomography (CT) or magnetic resonance imaging (MRI). For the use of microwave imaging devices like the ones used for breast tumor detection in clinical trials [19, 20], the number or radiating elements is important. Generally, it is desired to use high number of elements which should have unidirectional radiation patterns, while the cross-coupling among the radiating elements must be as small as possible. As a result, the use of compact unidirectional UWB radiators is considered. An effective solution is the use of cavity-backed slot antennas. Such an element is presented in **Figure 17**. An implementation of a microwave imaging device consisting of similar such radiators is described in [21, 30]. The front face can be made conformal to better match the non-planar surface, while the feeding cable can pass through the metallic cavity that is used to cancel the back radiation. The desired low profile of the cavity and the intermediate gap between the radiating antenna elements and the target make the design of such an antenna a challenging task. In order to avoid additional reflections, the target is immersed in a liquid with controlled dielectric constant since the UWB antennas' matching can be severely degraded when the antenna radiates in close proximity or in touch with the human body. The high effect of the human body with the complex electromagnetic characteristics on the UWB antenna performance necessitates the use of accurate human phantoms in the full-wave simulations. **Figure 17c** demonstrates a cavity-backed slot UWB antenna in close proximity with a realistic breast phantom in a setup used in a full-wave simulator in order to ensure the good performance of the antenna when it radiates in close proximity with human body parts. Although the metallic cavity increases the profile of the receiver, it is very useful since it blocks signals which are not coming directly from the target.

Different vendors [19, 20] use customized UWB antennas which serve better the preferred reconstruction algorithms that they use; however, the presented cavity-backed slot radiator is one of the best candidates for medical microwave devices.

8. Conclusion

Selected UWB antennas for personal area network communication systems, for positioning and location tracking, and for numerous radar applications are presented. It is evident that depending on the application, different antenna characteristics are required and different types of antennas are better suited. For wireless personal network devices, omnidirectional radiation patterns are preferred, and the compact size of the antenna with a low-profile planar design is the most useful, in most cases. There are several types of antennas that can be used; however, the monopole antenna is the most widely used design. Either CPW-fed or microstrip-fed UWB monopole is the most prominent solution. UWB monopoles have been used extensively for the implementation of reconfigurable UWB antennas which are capable of filtering out the interfering signals that share parts of the 3.1–10.6 GHz, FCC designated UWB spectrum. Frequency notch bands can be created with the addition of resonators which can be implemented on either the radiator or the feeding line or even the RF ground segments. Quarter-wavelength open stubs, half-wavelength linear segments, half-wavelength slots, and more complex in shape resonators, such as CLLs or SRRs, have been successfully used. For the reconfigurability attribution, RF switches are required, and depending on the resonator geometry and the biasing conditions, PIN diodes, FET switches, or MEMS switches can be used.

UWB antennas for positioning and tracking are used for both the interrogator and the RFID tags. Apparently different characteristics are needed for each case. For the interrogators high-gain, directive antennas with agile radiation patterns are preferred, while for the RFID tags, omnidirectional lightweight and low-cost antennas are needed. The RFIDs are often customized considering the electromagnetic characteristics of the items that they tag, and they can be either chipless or terminated with an IC load. The operation principle of chipless UWB RFIDs exploits the presence of a series of resonators which are coupled with the transmission line that connects the two antennas. One antenna receives the interrogator's signal, and the second one retransmits the modulated signal back to the reader. Chipless UWB RFIDs are entirely passive, and thus they can be easily fabricated using additive manufacturing technologies such as inkjet printing.

For radar applications, and depending on the target characteristics and the size and cost constraints, a wide variety of UWB antennas are used. For relatively long-range monostatic radars, such as ground-penetrating radars or through-wall imaging devices, high-gain directive antennas are used. Vivaldi antennas and their variations are the most common UWB antennas used for monostatic radars. For multistatic radars, like the ones used for microwave imaging for breast tumor detection, the requirements are very different. Since the target is relatively small and the size of the object under detection can be smaller than 1 cm^3, the radiators form a concave array. Cylindrical and hemispherical configurations are used, and the desired UWB antenna elements must have unidirectional radiation patterns, and they must be shielded from cross-coupling, while their size should be small to allow a large number of elements in a relatively small volume. Medical microwave imaging systems using UWB technology keep developing, and there are currently several vendors which have developed products which are cleared for clinical trials. Cavity-backed radiators are used covering different sub-bands in the UWB spectrum or even lower-frequency bands. These UWB antennas have conformal surfaces and are combined to form conformal arrays. They are co-designed with the human tissue target, since the radiation characteristics of UWB antennas change significantly when the antennas radiate in close proximity with the human body.

UWB antennas are integrated parts of devices which are used for a wide range of applications. For some applications, off-the-shelf UWB antennas can be successfully used or even antennas designed for similar scope, but for several other cases, UWB antennas must be customized and be co-designed with the surrounding environment since they can be easily mismatched and their assumed performance can be significantly degraded. Therefore, for the successful design of an effective UWB antenna, the antenna must be simulated and tested in a realistic environment that must consider the characteristics of the adjacent media and the surrounding objects.

Conflict of interest

All authors listed have contributed sufficiently to the project to be included as authors. The authors declare that there is no conflict of interest, in terms of financial or other regarding the publication of this book chapter.

Author details

Symeon Nikolaou* and Abdul Quddious
Frederick University and Frederick Research Center, Nicosia, Cyprus

*Address all correspondence to: s.nikolaou@frederick.ac.cy

References

[1] Nikolaou S, Abbasi MA. Design and development of a compact UWB monopole antenna with easily-controllable return loss. IEEE Transactions on Antennas and Propagation. 2017;**65**(4):2063-2067. DOI: 10.1109/TAP.2017.2670322

[2] Nikolaou S, Abbasi MA. Miniaturization of UWB antennas on organic material. International Journal of Antennas and Propagation. 2016;**2016**:12. DOI: 10.1155/2016/5949254. Article ID 5949254

[3] Tang MC, Wang H, Deng T, Ziolkowski RW. Compact planar ultrawideband antennas with continuously tunable, independent band-notched filters. IEEE Transactions on Antennas and Propagation. 2016;**64**(8):3292-3301. DOI: 10.1109/TAP.2016.2570254

[4] Ghahremani M, Ghobadi C, Nourinia J, Ellis MS, Alizadeh F, Mohammadi B. Miniaturised UWB antenna with dual-band rejection of WLAN/WiMAX using slitted EBG structure. IET Microwaves, Antennas and Propagation. 2018;**13**(3):360-366. DOI: 10.1049/iet-map.2018.5674

[5] Doddipalli S, Kothari A. Compact UWB antenna with integrated triple notch bands for WBAN applications. IEEE Access. 2019;7:183-190. DOI: 10.1109/ACCESS.2018.2885248

[6] Zhang J, Shen Z. Dual-band shared-aperture UHF/UWB RFID reader antenna of circular polarization. IEEE Transactions on Antennas and Propagation. 2018;**66**(8):3886-3893. DOI: 10.1109/TAP.2018.2839883

[7] Decarli N, Guidi F, Dardari D. Passive UWB RFID for tag localization: Architectures and design. IEEE Sensors Journal. 2016;**16**(5):1385-1397. DOI: 10.1109/JSEN.2015.2497373

[8] Liu J. Chipless UWB-RFID tag with spectral and temporal joint signatures. In: 2017 Sixth Asia-Pacific Conference on Antennas and Propagation (APCAP); 16 October 2017. pp. 1-3

[9] Galajda P, Pecovsky M, Gazda J, Drutarovsky M. Novel M-sequence UWB sensor for ground penetrating radar application. In: 2018 IEEE Asia-Pacific Conference on Antennas and Propagation (APCAP); 5 August 2018. pp. 110-111

[10] Liang J, Liu X, Liao K. Soil moisture retrieval using UWB echoes via fuzzy logic and machine learning. IEEE Internet of Things Journal. 2018;**5**(5):3344-3352. DOI: 10.1109/JIOT.2017.2760338

[11] Qi F, Liang F, Liu M, Lv H, Wang P, Xue H, et al. Position-information-indexed classifier for improved through-wall detection and classification of human activities using UWB bio-radar. IEEE Antennas and Wireless Propagation Letters. 2019;**18**(3):437-441. DOI: 10.1109/LAWP.2019.2893358

[12] Song Y, Hu J, Chu N, Jin T, Zhang J, Zhou Z. Building layout reconstruction in concealed human target sensing via UWB MIMO through-wall imaging radar. IEEE Geoscience and Remote Sensing Letters. 2018;**15**(8):1199-1203. DOI: 10.1109/LGRS.2018.2834501

[13] Shao W, Edalati A, McCollough TR, McCollough WJ. A time-domain measurement system for UWB microwave imaging. IEEE Transactions on Microwave Theory and Techniques. 2018;**66**(5):2265-2275. DOI: 10.1109/TMTT.2018.2801862

[14] Wang F. Assembly conformal antenna array for wearable microwave breast imaging application. In: Loughborough Antennas &

Propagation Conference (LAPC 2017); Loughborough; 2017. pp. 1-5

[15] Wang F, Arslan T, Wang G. Breast cancer detection with microwave imaging system using wearable conformal antenna arrays. In: 2017 IEEE International Conference on Imaging Systems and Techniques (IST); 18 October 2017. pp. 1-6

[16] Mukherjee S, Udpa L, Udpa S, Rothwell EJ, Deng Y. A Time reversal-based microwave imaging system for detection of breast tumors. IEEE Transactions on Microwave Theory and Techniques. May 2019;**67**(5):2062-2075. DOI: 10.1109/TMTT.2019.2902555

[17] Song H, Sasada S, Masumoto N, Kadoya T, Shiroma N, Orita M, et al. Detectability of breast tumors in excised breast tissues of Total mastectomy by IR-UWB-radar-based breast cancer imaging. IEEE Transactions on Biomedical Engineering. 17 Dec 2018: (Early access). DOI: 10.1109/TBME.2018.2887083

[18] Bourqui J, Sill JM, Fear EC. A prototype system for measuring microwave frequency reflections from the breast. Journal of Biomedical Imaging. 2012;**2012**:9. DOI: 10.1155/2012/851234

[19] Translational Medical Device Lab: Imaging and Diagnostics [Internet]. Available from: https://tmdlab.ie/ [Accessed: 24 April 2019]

[20] Micrima: Developing Technology for Breast Cancer Screening [Internet] Available from: https://micrima.com/ [Accessed: 24 April 2019]

[21] Porter E, Bahrami H, Santorelli A, Gosselin B, Rusch LA, Popović M. A wearable microwave antenna array for time-domain breast tumor screening. IEEE Transactions on Medical Imaging. 2016;**35**(6):1501-1509. DOI: 10.1109/TMI.2016.2518489

[22] Nikolaou S. Design and implementation of compact reconfigurable antennas for UWB and WLAN applications [Doctoral dissertation]. Georgia Institute of Technology; 2006

[23] Nikolaou S, Davidović M, Nikolić M, Vryonides P. Triple notch UWB antenna controlled by three types of resonators. In: 2011 IEEE International Symposium on Antennas and Propagation (APSURSI); 3 July 2011. pp. 1478-1481. DOI: 10.1109/APS.2011.5996574

[24] Quddious A, Ali M, Abbasi B, Vryonides P, Nikolaou S, Antoniades MA, et al. Reconfigurable notch-band UWB Antenna with RF-to-DC rectifier for dynamic reconfigurability. In: 2018 IEEE International Symposium on Antennas and Propagation & USNC/URSI National Radio Science Meeting; 8 July 2018. pp. 283-284. DOI: 10.1109/APUSNCURSINRSM.2018.860868

[25] Nikolaou S, Kingsley ND, Ponchak GE, Papapolymerou J, Tentzeris MM. UWB elliptical monopoles with a reconfigurable band notch using MEMS switches actuated without bias lines. IEEE Transactions on Antennas and Propagation. 2009;**57**(8):2242-2251. DOI: 10.1109/TAP.2009.2024450

[26] Zhang C, Hu Y, Jin X, Huang X. High performance linearly tapered slot antenna (ltsa) using parasitic patch. In: 2017 Sixth Asia-Pacific Conference on Antennas and Propagation (APCAP); 16 October 2017. pp. 1-3

[27] Zhang F, Zhang FS, Zhao G, Lin C, Jiao YC. A loaded wideband linearly tapered slot antenna with broad beamwidth. IEEE Antennas and Wireless Propagation Letters. 2011;**10**:79-82. DOI: 10.1109/LAWP.2011.2106477

[28] Nikolaou S, Ponchak GE, Papapolymerou J, Tentzeris MM.

Conformal double exponentially tapered slot antenna (DETSA) on LCP for UWB applications. IEEE Transactions on Antennas and Propagation. 2006;**54**(6):1663-1669. DOI: 10.1109/TAP.2006.875915

[29] Nikolaou S, Tentzeris MM, Papapolymerou J. Study of a conformal UWB elliptical monopole antenna on flexible organic substrate mounted on cylindrical surfaces. In: 2007 IEEE 18th International Symposium on Personal, Indoor and Mobile Radio Communications; 3 September 2007. pp. 1-4

[30] Gibbins D, Klemm M, Craddock I, Preece A, Leendertz J, Benjamin R. Design of a UWB wide-slot antenna and a hemispherical array for breast imaging. In: 2009 3rd European Conference on Antennas and Propagation; 23 March 2009. pp. 2967-2970

Chapter 5

Review on UWB Bandpass Filters

Li-Tian Wang, Yang Xiong and Ming He

Abstract

Rapid development of a number of wireless communication systems imposed an urgent requirement for a technology which contains multi-wireless communication standard. Since the ultra-wideband (UWB) technologies are of advantage in broad bandwidth and high-speed transmission, much attention has been paid to exploiting the UWB bandpass filters. In this chapter, the development process of the UWB bandpass filters and the regulation of the UWB bandpass filter are initially introduced. Subsequently, the application scenarios of UWB filters in UWB communication systems and unique merits of UWB filters were explored. In addition, the primary performance specifications of the UWB filters, including insertion loss, return loss, the level of out-of-band attenuation, and roll-off rate, are also presented. After a brief discussion of microwave network theory, several methods for implementing UWB filters are summarized. Furthermore, the design of the UWB filter with notch band is presented in Section 5. The last section, the Conclusion section, is given at the end of this chapter.

Keywords: UWB bandpass filter, multimode resonator, step-impedance resonator, wide stopband, high selectivity, notch band

1. Introduction

The ultra-wideband (UWB) communication technology with a long history is developed rapidly in the past few decades. Since 1989, the UWB was first employed by the Defense Advanced Research Projects Agency (DAPRA) as a term, and the DAPRA also proposed the bandwidth definition of the UWB. In fact, the UWB technology was only authorized to be applied in military communications. Since February 2002, the development of UWB has undergone a great change. The Federal Communications Commission (FCC) finally released the UWB spectrum globally for data communication or radar and security field for civilian application and redefined the bandwidth of UWB, which specifies that the UWB radio-frequency signal has a fractional bandwidth (FBW) greater than 20% or 10 dB absolute bandwidth greater than 500 MHz. According to the definition of FCC part 15 [1], the authorized band allocated to the UWB communication systems is ranging from 3.1 to 10.6 GHz. Unprecedented 7.5 GHz of bandwidth is the largest bandwidth of any commercial terrestrial system has ever allocated. The 3 dB FBW of the UWB can reach 109%, and FCC emission mask specified that the transmission power does not exceed −41.3 dBm/MHz (75 nW/MHz). The way of sharing the spectrum with extremely low-power spectral densities (PSD) is of paramount significance in present intense crowned spectrum circumstance. The major merits of the UWB are as follows:

Firstly, high date rate: according to the Shannon formula for channel capacity [2], the maximum infallibility information date rate of the system in the additive white Gaussian noise (AWGN) channel can be expressed as

$$C = B\log_2\left(1+\frac{S}{N}\right) \quad (1)$$

where B stands for channel width and S/N denotes the signal-to-noise ratio. Hence, it can be concluded from Eq. (1) that even if the signal-to-noise ratio values are as low as 0.1 (−10 dB), the system's data rate still can reach as high as 1 Gbps. It fully demonstrates the extremely high date rate of the UWB system. Secondly, strong anti-interference ability: UWB resorts carrierless communication with nanosecond pulses. With Fourier transform, it can be derived that the power spectral density is dramatically wide with low-energy density, which reveals that the UWB system is of excellent concealment. Thirdly, high resolution ratio of time and space: the UWB is operating at high frequency with a nanosecond resolution of time domain, and the short wavelength at the RF enables spatial resolution of 0.1 m approximately. The rapid development of 5G [3] and the Internet of Things (IoT) [4] has an urgent demand for high response speed and high positioning accuracy, and the UWB can perfectly meet this requirement. The emergence of key reports and research process, whether from an academic or engineering perspective, has greatly advanced the development of UWB over the past few decades.

The UWB bandpass filters served as key building block in UWB wireless communication systems to regulate the FCC UWB masks have aroused much research interest in this century. And various attempts to design UWB have been reported continuously. The UWB bandpass with a FBW of more than 20% have been reported with simple design methodology and excellent passband performance since 2012 [5, 6]. However, for the FCC authorized specification, 109% of the FBW is actually an unprecedented challenge in approaching UWB bandpass filters design. Despite the well-established comprehensive design theory for narrowband bandpass filters with varied specification [7–10], the synthesis design methods for UWB bandpass filters are not suitable to employ existing powerful design theory foundations.

Various techniques have been presented to develop the UWB bandpass filters. One of the straightforward methods is cascading a low-pass filter and a high-pass filter to accomplish UWB bandpass filter [11–13]. Though considerable wideband is realized in [11], the occupied circuit size needs to be further reduced. To achieve UWB bandpass filters with compact size and simple design process, multimode resonator (MMR) has been presented [14–25]. In [16], the UWB bandpass filter is achieved with wide stopband, and 40 dB attenuation can be realized within frequency ranging from 12.0 to 16.0 GHz. In [19], quintuple-mode resonator is introduced to design UWB bandpass filter and sharp shirt, and wide upper stopband is achieved simultaneously. A UWB bandpass filter with 20 dB out-of-band suppression up to 25.1 GHz is proposed in [21].In [25], a novel MMR with interdigital-coupled-microstrip line sections is implemented, which can excite seven transmission poles to design UWB bandpass filter with high roll-off rate. In summary, design of UWB bandpass filters by using MMR is of compact size and with multi-transmission poles, whereas the range of out-of-band rejection is still insufficient since harmonic effects. Similar to the MMR, the stub-loaded multimode resonator (SLMMR) is another ideal structure to design UWB bandpass filters owing to its simple structure and easy design procedure [26–33]. In [26], a highly selective UWB bandpass filter is achieved by short-circuit stub-loaded structure, which can excite 11 resonant modes to fulfill the requirement of UWB with miniature circuit size.

The stub-loaded quintuple-mode resonator is employed to design UWB bandpass filter with two transmission zeros near the lower and the upper cutoff frequencies in [30]. To address the issue of harmonic effect to obtain wide stopband, the step-impedance resonator (SIR) is utilized to design UWB bandpass filters with remoted harmonic [33–37]. A UWB bandpass filter with more than 30 GHz out-of-band attenuation is approached by using SIR in [34]. The novel ring resonators are considered as an effective way to design UWB bandpass filters attributed to its miniature size and multiple resonance behavior [38–46]. In [39], a design of UWB bandpass filter with extremely compact circuit size (0.46 cm^2, without feedlines) by using quintuple ring resonator is proposed. In [42], UWB with switchable bandwidth is also investigated by implementing a ring resonator, and tunable passband ratio of 1.22:1.13:1 is obtained. Another major category of UWB bandpass with desired UWB passband performance is based on the parallel-coupled lines [47, 48]. In [48], by using parallel-coupled microstrip lines, a UWB bandpass filter with a passband from 3.1 to 10.6 GHz of less than 1 dB insertion loss is accomplished; meanwhile, the attenuation level can reach 40 dB in stopband. In order to cater for the urge demand for miniaturization, UWB bandpass filters with multilayer structures have been extensively investigated and reported [49–58]. In [57], design of an eight-pole UWB filter is demonstrated; meanwhile the proposed UWB filter not only has merits of miniature circuit size but also processes a 38.1 dB out-of-band suppression by utilizing the multilayer structure. In addition to the aforementioned techniques, there were also other routines to obtain the UWB bandpass filter, such as semi-lumped UWB bandpass filter [58, 59] and UWB bandpass filter designed with right-/left-banded structure [60–62]. Furthermore, for the purpose of achieving the UWB communication while eliminating other inferences of current communication systems, notch band UWB bandpass filter is presented [60, 63–77] and will be demonstrated in detail in Section 6 of this chapter.

This chapter mainly focuses on the various approaches to achieve UWB bandpass filter and the discussion of several conventional methods for high-performance UWB filter with wide stopband, high out-of-band attenuation, sharp selectivity, and miniaturization. Therefore, the organization of this chapter is as follows: in Section 2, application scenarios of UWB, development history of UWB, and the UWB regulations established by the FCC are briefly demonstrated. In Section 3, the major specifications of the UWB filters as well as the foundation of design methodology are illustrated. Section 4, the key section, focuses on varied approaches to realize the UWB filter design. Common ways for accomplishing the design of UWB filters can be classified into the following categories: one of the general methods of designing UWB bandpass filter is using multimode resonator (MMR) (Section 4.1), and similar to the method in Section 4.1, UWB bandpass filters are also realized by using a stub-loaded multimode resonator (SLMMR) (Section 4.2). The methods of implementing the UWB bandpass filter with multi-layer structure, parallel-coupled line, and step-impedance resonator design methodology are, respectively, reviewed in Sections 4.3–4.5. In order to fulfill the requirement to eliminate the RF interference within the UWB band, the UWB bandpass filter with notch band has been designed and reported extensively, which is reviewed in Section 5. Section 6, the Conclusion section, will be given at the end of this chapter.

2. Regulation and application

The UWB wireless communication has been only authorized to the military communication for 42 years. Since 2002, the FCC released the unlicensed employment of UWB for commercial purpose and classified the application scenario into imaging system, vehicle radar system, and communication and measurement system. According to the regulation of FCC, the f_L and f_H are located at 3.1 and 10.6 GHz, where f_L and f_H, respectively, stand for the frequencies with 10 dB attenuation of the upper and lower sidebands. Therefore, the center frequency of UWB bandpass filter is expressed as

$$f_C = \frac{f_H + f_L}{2} = 6.85 \text{ GHz} \tag{2}$$

Meanwhile, the *FBW* can also derived as

$$FBW = \frac{2(f_H - f_L)}{f_H + f_L} = 109.5\% \tag{3}$$

For the purpose of evading the interference of UWB systems and existing communication systems (such as GPRS, WLAN, TD-LTE, and mobile cellular), the radiation spectral density of UWB systems is strictly limited and regulated, the highest power spectral density of the UWB systems regulated not exceeding −41.3 dBm/MHz, as illustrated in **Figure 1**. It is worth noting that the regulation also varies depending on the indoor and outdoor circumstance.

As depicted in **Figure 2**, the UWB band has an extremely high FBW and unparalleled 7.5 GHz absolute bandwidth; the UWB therefore has potential in many applications. The UWB can be applied to support large channel capacity since its huge bandwidth, whereas its propagation distance is limited by low effective isotropic radiated power (EIRP). Therefore, the UWB is an ideal candidate for short-distance high-rate communication. For detecting, the UWB has a dramatic penetrating ability by using its outstanding weak narrow pulse of baseband, which can easily penetrate the leaves, the earth's surface, the clouds, and the concrete; even objects behind the obstacle can also detected. For locating, high positioning accuracy can be accomplished by UWB technology, whether for military or civilian application.

The UWB bandpass filters have the responsibility to remove the unwanted signals and noise in UWB communication system. As shown in **Figure 3**, for the transmitting system, the modulated signal is directly filtered by using the UWB bandpass filter, and the UWB bandpass filter is also a critical component in RF front

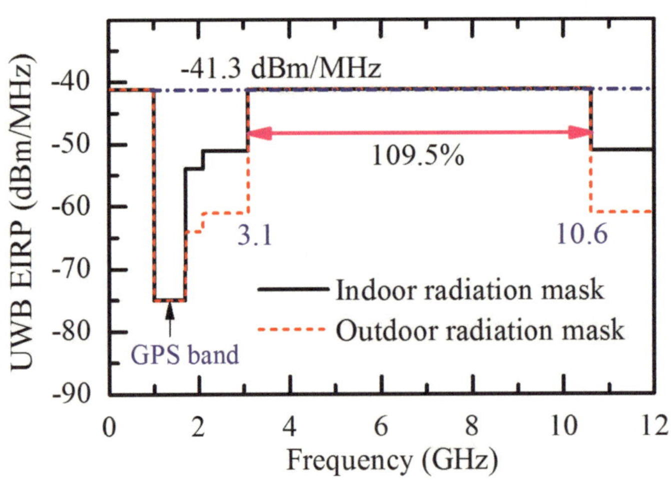

Figure 1.
Indoor radiation masks and outdoor radiation mask regulated for UWB system by FCC.

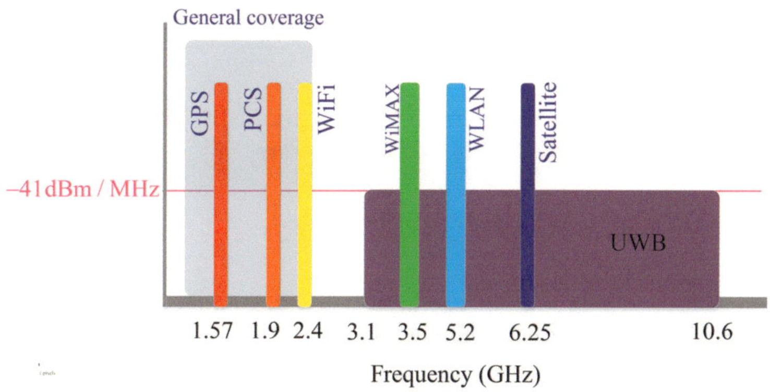

Figure 2.
The comparison between the UWB spectrum and spectrum of currently commercial communication systems.

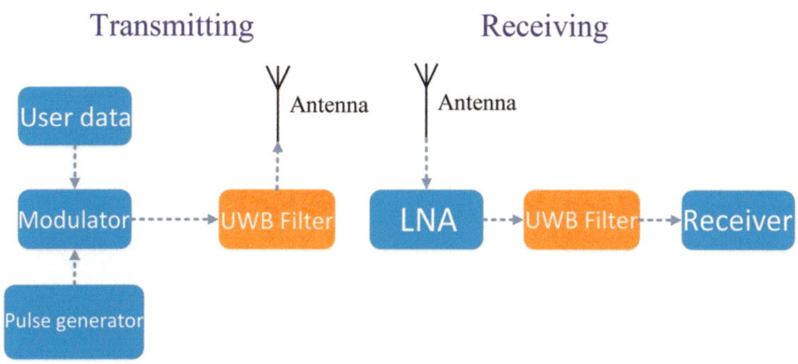

Figure 3.
The sketch of transmitting system and receiving system with primary components.

end of the receiving system [5]. Therefore the UWB bandpass filters are known as an inevitable key building block of wideband communication systems, and the filtering performance of UWB bandpass filter will directly related to the performance of the entire UWB system.

3. Microwave foundation of designing UWB bandpass filters

3.1 Performance specifications

The performance is a critical factor for UWB filters even in any engineering device, and the major parameters for UWB filter performance evaluating are as follows:

1. Insertion loss (IL): insertion loss is the attenuation caused by the introduction of the device between the in port and out port, usually expressed in dB. The insertion loss can be calculated as follows:

$$IL = 10 \log \frac{P_{in}}{P_{out}} \text{ (dB)} \tag{4}$$

where P_{in} is the input transmitted power and the P_{out} is the output received power. In addition to the mismatching loss, the actual bandpass filters have a series of other losses. Firstly, dielectric loss, can expressed as

$$\alpha_d' = 27.3\frac{\tan\delta}{\lambda_g}\ \mathrm{dB/cm} \tag{5}$$

where the λ_g is the guide wavelength of 50 Ω microstrip line at frequency f. Secondly, the conductor loss, can be derived from

$$\alpha_a = \frac{\sqrt{\pi f \mu_0 \sigma}}{4(w+t)\sigma \cdot Z_0} \tag{6}$$

where μ_0 is the permeability of vacuum, σ is the conductivity, w is the width of conductor, and t is the conductor thickness.

Thirdly, the dielectric loss, can be written as

$$\alpha_d = \tan\delta \cdot \pi/\lambda_g \tag{7}$$

2. Return loss (RL): return loss is the ratio of the reflected power to the incident wave power, expressed in dB:

$$RL = 10\log\frac{P_{re}}{P_{in}}\ (\mathrm{dB}) \tag{8}$$

3. FBW and center frequency

4. Roll-off rate (ROR): the ROR is a critical specification for evaluating passband selectivity and can be defined as follows:

$$ROR = \frac{|\delta_{-20dB} - \delta_{-3dB}|}{|f_{-20dB} - f_{-3dB}|} \tag{9}$$

where $\delta_{-20\ \mathrm{dB}}$ and $\delta_{-3\ \mathrm{dB}}$ are attenuation point at −20 and −3 dB, respectively. $f_{-20\ \mathrm{dB}}$ and $f_{-3\ \mathrm{dB}}$ are, respectively, −20 and −3 dB stopband frequency.

5. Group delay: the ratio of phase variation to frequency variation is utilized to describe the overall delay of signal though the device. The group delay can be derive as

$$\tau = -\frac{\partial\varphi_{21}(\omega)}{\partial\omega} \tag{10}$$

6. Out-of-band suppression level: the stopband suppression level is applied to evaluate the out-of-band performance of the UWB bandpass filter.

7. Upper stopband bandwidth: it is worth noting that there is no spike in the stop band.

8. Transmission poles: multi-transmission poles prone to achieve UWB bandpass filters with sharp shirt.

9. Transmission zeros: the UWB bandpass filter with multi-transmission zeros tends to process excellent out-of-band rejection and high selectivity of passband.

3.2 Foundation of conventional transmission line filter analysis

The critical step in the design of a conventional transmission line UWB filter is to select the appropriate electrical lengths/impedances of transmission lines to adjust the resonant modes to fulfill the design specifications of UWB bandpass filter. For the purpose of establishing the expression of the resonant frequencies and each electrical lengths or impedances, the Y_{in} of UWB filter needs to be derived, and the resonance condition can be calculated by using the following expression:

$$\text{Im}\,(Y_{\text{in}}) = 0 \tag{11}$$

As demonstrated in **Figure 4**, the detailed steps for solving resonant frequencies with numerical calculation are as follows:

Step 1, Initialization. The electrical lengths θ_i (i = 1, 2, ..., n), f_0 (reference frequency for electrical length calculation), and frequency sweep range should be given.

Step 2, Recalculating the electrical lengths. When the frequency f_i is considered, all of the electrical lengths should be recalculated as $\theta_n{}' = \theta_n f_i / f_0$. Then, we substitute the updated electrical lengths into Eqs. (2)–(8).

Step 3, if Eq. (1) is satisfied, that is, $f_1 = f_i$ is the first resonant frequency that we are searching for, then the resonant frequency f_1 should be saved and turn to the next step. If Eq. (1) is not satisfied, the program turns to the next step directly.

Step 4, considering the next frequency f_{i++}.

Step 5, is the new value of f_{i++} beyond the frequency sweep range? If the answer is yes, then quit and end the program. If the answer is no, then go to step 2.

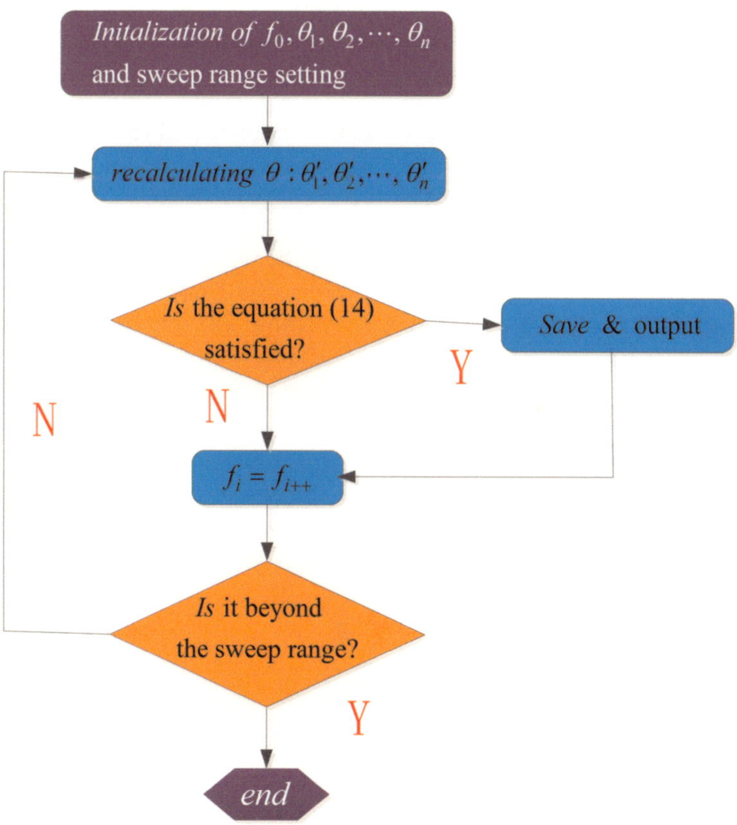

Figure 4.
Flow chart of solving the resonate frequency.

3.2.1 Classical even-odd-mode analysis method

Since the odd-mode resonant frequencies of the symmetrical structure are orthogonal to the even-mode resonant frequencies, the whole transmission line model can be divided into odd-mode and even-mode circuits. Therefore, the resonant frequencies are then derived separately, which dramatically reduces the computation of resonant modes. It is worth noting that with odd-mode excitation, the symmetrical planes are considered to be grounded and with even-mode excitation, the symmetrical planes are considered to be open.

Even-/odd-mode input admittance can be obtained from the even–/odd-mode equivalent circuit, and Eq. (11) can be replaced by the following equations:

$$\mathrm{Im}\left(Y_{\mathrm{ine}}\right) = 0 \tag{12}$$

$$\mathrm{Im}\left(Y_{\mathrm{ino}}\right) = 0 \tag{13}$$

3.2.2 Classical [ABCD] matrix analysis method

The analysis of traditional transmission line filters with asymmetric structures is no longer within the application scope of classical odd-even-mode analysis method. To overcome this issue, the ABCD matrix method is employed to approach the overall transmission ABCD matrix; the Y_{in} is then derived from the ABCD matrix of the overall structure. The ABCD matrix of several typical transmission line models and the ABCD matrix of several conventional circuit elements are depicted in **Figure 5**.

3.2.3 Analysis of parallel-coupled lines

The analysis of parallel-coupled lines is more complicated than that of series/shunt transmission lines. One of the reliable ways is to analyze the parallel-coupled lines as a four-port component, and parameters of parallel-coupled lines are shown in **Figure 6**.

Different paralleled coupling conditions and the position of the in/out port correspond to varied initial conditions. Therefore, the Z matrix of parallel-coupled lines can be solved according to this initial condition. The four-port impedance matrix is given as follows.

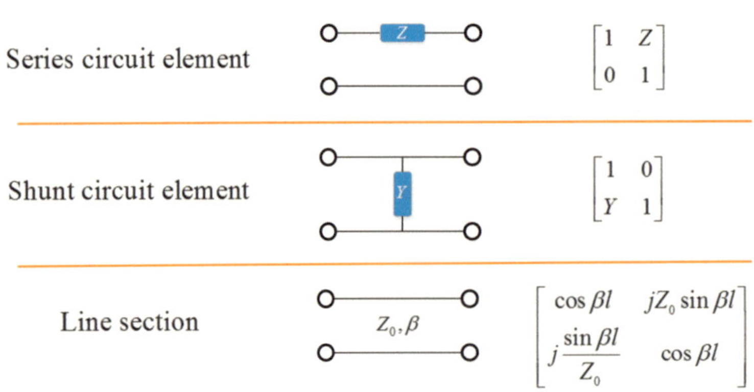

$$\begin{bmatrix} 1 & Z \\ 0 & 1 \end{bmatrix} \qquad \begin{bmatrix} 1 & 0 \\ Y & 1 \end{bmatrix} \qquad \begin{bmatrix} \cos\beta l & jZ_0 \sin\beta l \\ j\dfrac{\sin\beta l}{Z_0} & \cos\beta l \end{bmatrix}$$

Figure 5.
Classical transmission line structure and their ABCD matrix.

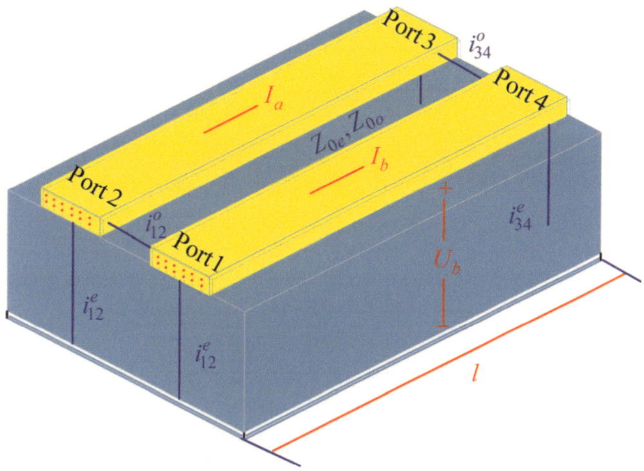

Figure 6.
Electrical diagram of parallel-coupled line.

$$\begin{bmatrix} U_1 \\ U_2 \\ U_3 \\ U_4 \end{bmatrix} = \begin{bmatrix} Z_{11} & Z_{12} & Z_{13} & Z_{14} \\ Z_{21} & Z_{22} & Z_{23} & Z_{24} \\ Z_{31} & Z_{32} & Z_{33} & Z_{34} \\ Z_{41} & Z_{42} & Z_{43} & Z_{44} \end{bmatrix} \begin{bmatrix} I_1 \\ I_2 \\ I_3 \\ I_4 \end{bmatrix} \tag{14}$$

where

$$Z_{11} = Z_{22} = Z_{33} = Z_{44} = -\mathrm{j}\frac{Z_{0e} + Z_{0o}}{2} \cot\theta \tag{15}$$

$$Z_{12} = Z_{21} = Z_{34} = Z_{43} = -\mathrm{j}\frac{Z_{0e} - Z_{0o}}{2} \cot\theta \tag{16}$$

$$Z_{13} = Z_{31} = Z_{24} = Z_{42} = -\mathrm{j}\frac{Z_{0e} - Z_{0o}}{2} \csc\theta \tag{17}$$

$$Z_{14} = Z_{41} = Z_{23} = Z_{32} = -\mathrm{j}\frac{Z_{0e} + Z_{0o}}{2} \csc\theta \tag{18}$$

4. Common UWB bandpass filters

4.1 UWB bandpass filters using MMR

A quintuple-mode resonator is proposed to design UWB bandpass filter, and the physical layout of the presented UWB filter is sketched in **Figure 7** [19]. Since the whole structure is symmetrical along the *T–T'* line, classical odd-even-mode method is adopted to analyze the quintuple-mode resonator. As demonstrated in **Figure 8**, five resonant modes can be generated by quintuple-mode resonator; besides, owing to the loaded stub, two transmission zeros are realized both at lower and upper cutoff frequencies; thus, high selectivity is approached. As shown in **Figure 9**, the measurement results are in good agreement which shows sharp skirt and ultra-wide stopband of the UWB bandpass filter (**Figure 9**).

4.2 UWB bandpass filters using SLMMR

As illustrated in **Figure 10**, dual short stub-loaded resonator is presented to construct UWB transmission characteristics [31]. Owing to symmetrical structure

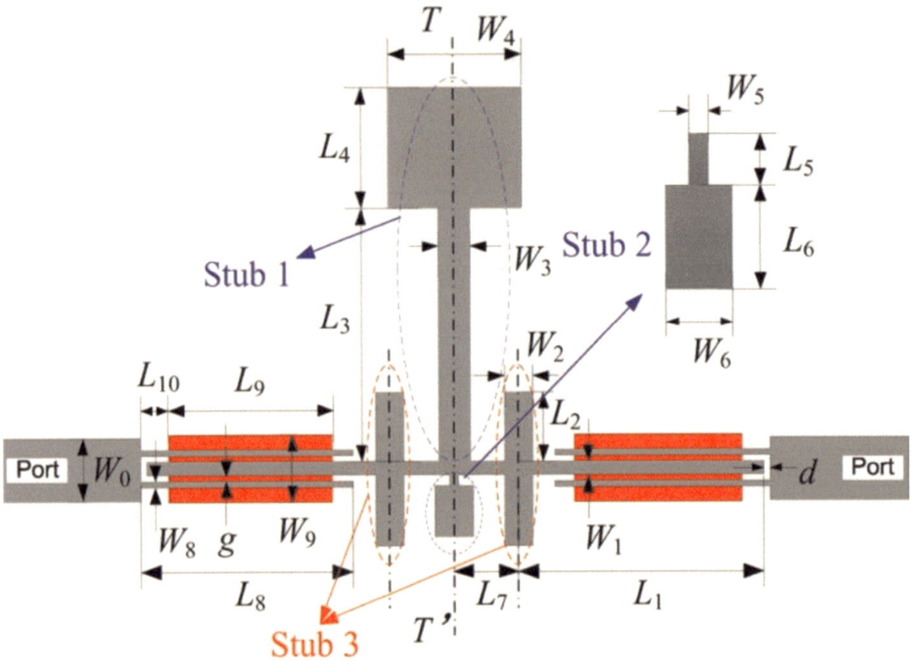

Figure 7.
Physical layout of the quintuple-mode resonator.

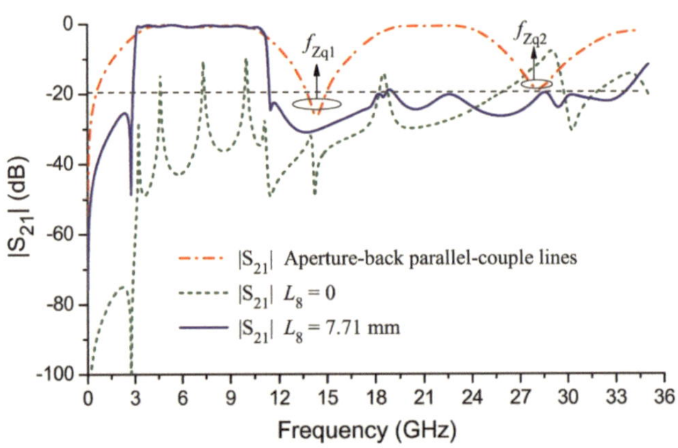

Figure 8.
Transmission coefficient $|S_{21}|$ *versus weak coupling and strong coupling.*

of the presented SLMMR, classical odd-even-mode method can be introduced to analyze the resonant modes of UWB filter. With even-mode excitation and odd-mode excitation, the input admittance can be, respectively, written as follows:

$$Y_{ine} = Y_c \frac{(jY_1 \tan\theta_1 - jY_2 \cot\theta_2 + jY_3 \tan\theta_3) + jY_c \tan\theta_c}{Y_c + (-Y_1 \tan\theta_1 + Y_2 \cot\theta_2 - Y_3 \tan\theta_3) \tan\theta_c} \tag{19}$$

$$Y_{ino} = Y_c \frac{(-jY_1 \cot\theta_1 - jY_2 \cot\theta_2 + jY_3 \tan\theta_3) + jY_c \tan\theta_c}{Y_c + (Y_1 \cot\theta_1 + Y_2 \cot\theta_2 - jY_3 \tan\theta_3) \tan\theta_c} \tag{20}$$

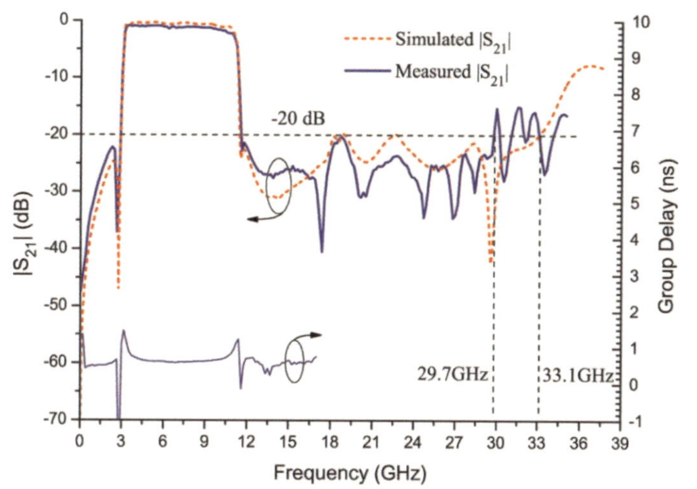

Figure 9.
Simulated and measurement frequency responses of presented UWB.

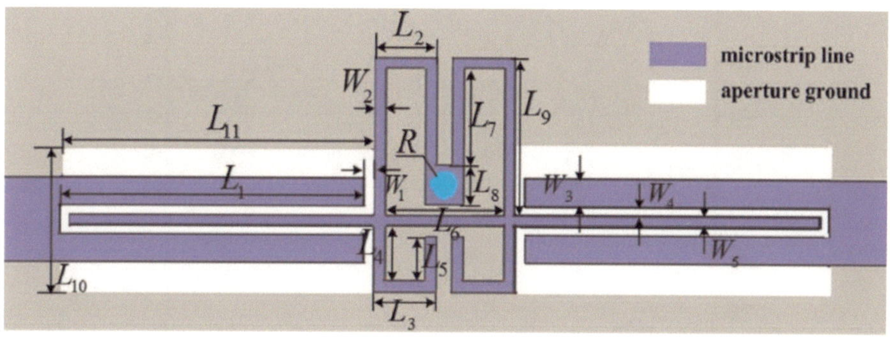

Figure 10.
Schematic diagram of proposed UWB bandpass filter with SLMMR.

By using the numerical calculation method mentioned in **Figure 4**, the design graphs for implementing UWB bandpass filter are sketched in **Figure 11**. For example, by properly choosing the values of θ_2, the excited four resonant frequencies can be easily adjusted to the desired UWB specifications. Therefore, the first four resonant modes are located at 2.86, 5.58, 8.56, and 10.21 GHz, and the dimension parameters are optimized by IE3D as follows: $L_1 = 8$, $L_2 = 1.6$, $L_3 = 1.6$ $L_4 = 1.4$, $L_5 = 1.2$, $L_6 = 3$, $L_7 = 2.5$, $L_8 = 1$, $L_9 = 4$, $L_{10} = 3.6$, $L_{11} = 8$, $W_1 = 0.2$, $W_2 = 0.3$, $W_3 = 0.6$, $W_4 = 0.2$, and $W_5 = 0.3$. It can be observed in **Figure 12** that simulation results are in good agreement with measurement results, which shows UWB bandpass characteristics with small and flat group delay in the passband.

4.3 UWB bandpass filters with multilayer structure

In order to design bandpass filter with UWB performance while occupied compact size, dual-layered structure is proposed in [56]. The UWB filter is constructed by substrate integrate waveguide (SIW) ridge resonator, and the bandwidth of the UWB filter can be easily tuned by properly changing the width of rod in ridge resonator. The scheme diagram is sketched in **Figure 13**.

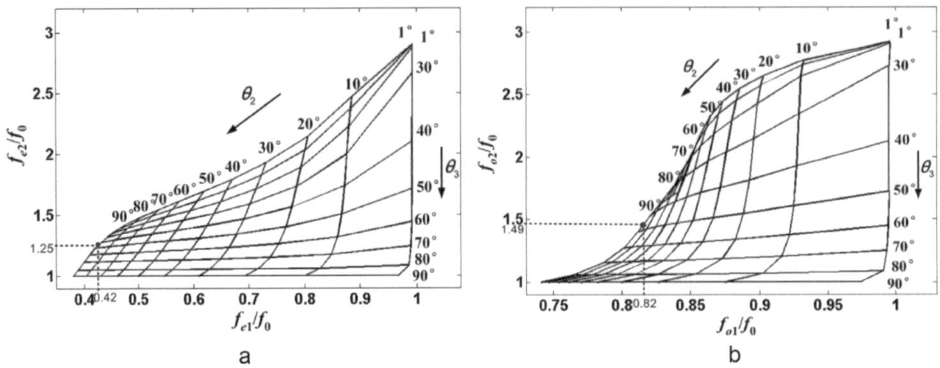

Figure 11.
Design graph for SLMMR, (a) normalized even-mode resonance frequencies versus θ_2, (b) normalized odd-mode resonance frequencies versus θ_2.

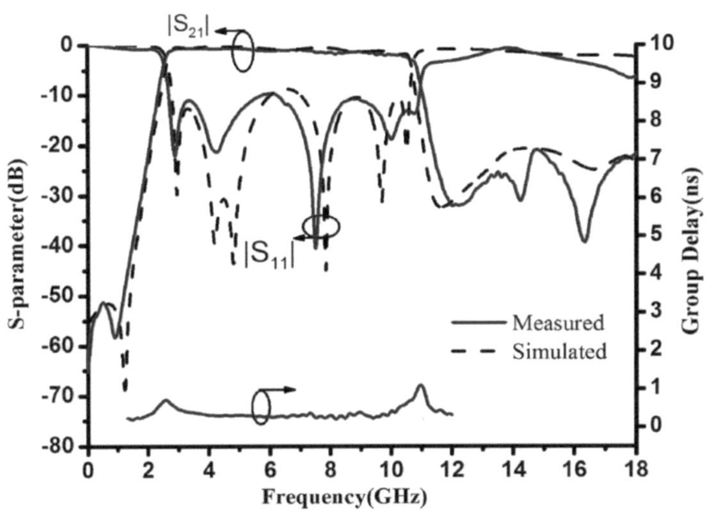

Figure 12.
Measured results versus simulated results of fabricated UWB filter.

As depicted in **Figures 14** and **15**, the bandwidth of UWB filter increases as the R_s increases, and the coupling strength and the bandwidth are both decreased as R_L lessens; thus, the bandwidth of UWB filter can be easily tuned by properly adjusting the R_s and R_L, and design parameters are finally chosen as $W_0 = 0.4$, $L_0 = 4$, $W_1 = 0.55$, $L_1 = 4.85$, $W_2 = 0.85$, $L_2 = 5.1$, $W_3 = 1.4$, $L_3 = 4.9$, $W_4 = 2.25$, $L_4 = 4.74$, $W_5 = 3.15$, and $L_5 = 4.1$. For the purpose of validating the design methodology, the dual-layer UWB bandpass filter is fabricated on the substrate of Rogers 6006 with relative permittivity if 6.15 and measured. The measurement results indicate that the proposed UWB filter is of extremely low insertion loss (<1 dB) and 47 dB stopband suppression up to 17.4 GHz with compact size, which can be observed in **Figure 16**.

4.4 UWB bandpass filters with parallel-coupled lines

The parallel-coupled lines can also employ to design UWB bandpass filter with simple structure. In [47], shorted coupled line structure and $\lambda/4$ shorted stub are introduced to achieve UWB bandpass filter with compact size. The ideal transmission line model of proposed UWB bandpass filter is demonstrated in **Figure 17**.

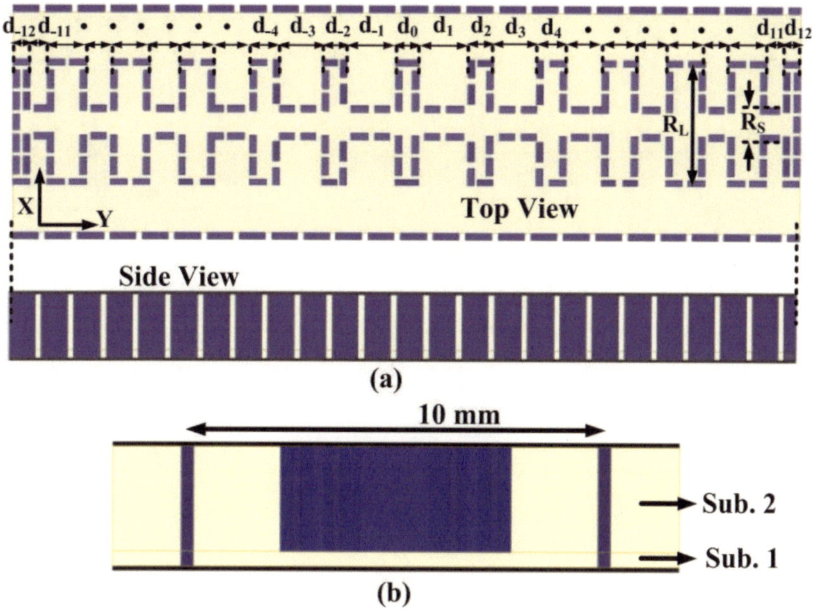

Figure 13.
Physical layout of the presented UWB filter. (a) Top band side view. (b) Front view.

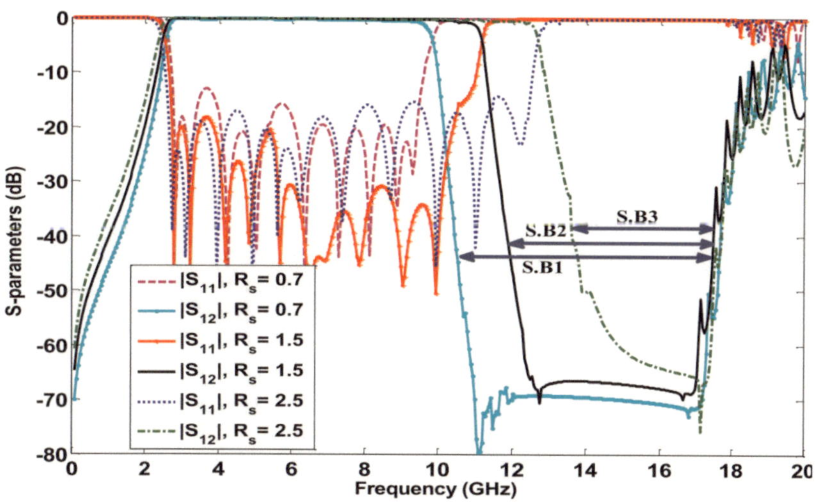

Figure 14.
Different transmission characteristics versus varied R_s.

Since the shorted coupled line structure is not a symmetrical structure, the ABCD matrix analysis method is employed to solve the input admittance of the proposed UWB bandpass filter, and the Y-matrix of this filter can be written as

$$Y = Y_{upper} + Y_{lower} \tag{21}$$

where

$$Y_{upper} = \begin{bmatrix} \frac{D}{B} & \frac{BC - AD}{B} \\ -\frac{1}{B} & \frac{A}{B} \end{bmatrix} \tag{22}$$

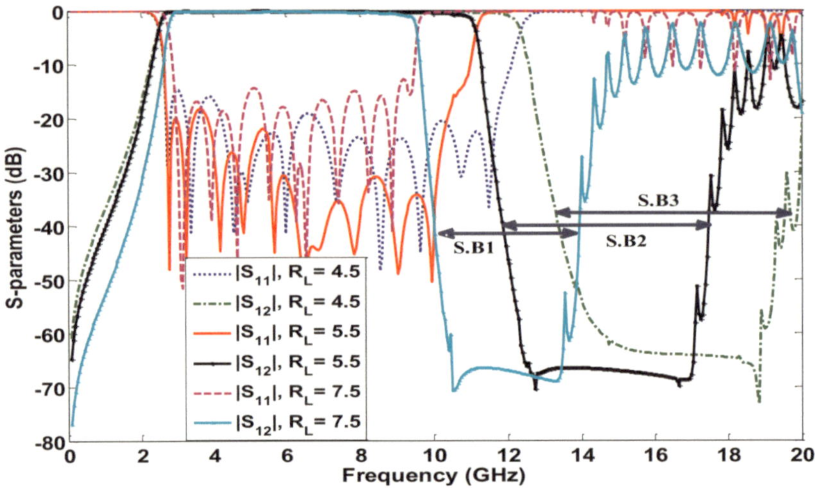

Figure 15.
Variation of frequency responses against varied R_L.

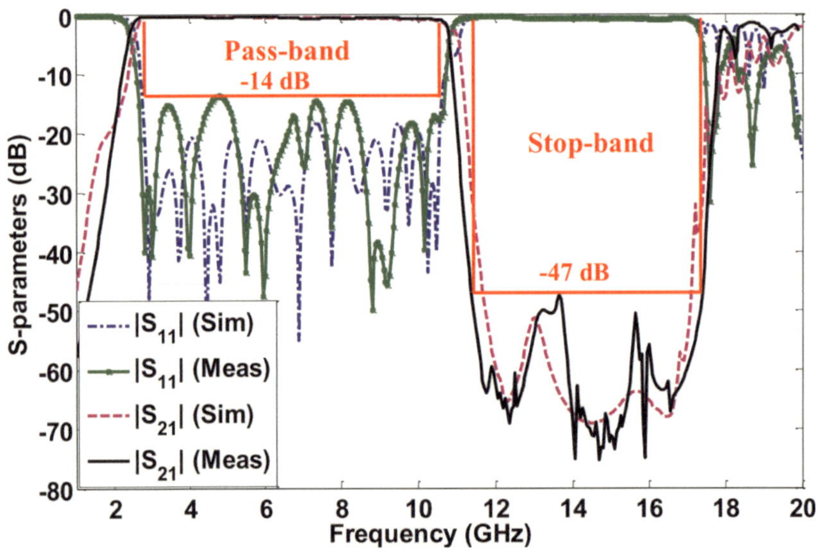

Figure 16.
Measurement results and simulated results of fabricated UWB filter.

where the whole ABCD matrix can be derived by

$$\begin{bmatrix} A & B \\ C & D \end{bmatrix} = M_2 M_3 M_4 M_5 M_6 \tag{23}$$

where

$$M_n = \begin{bmatrix} \cos\theta_n & jZ_n \sin\theta_n \\ j\left(\frac{1}{Z_n}\right)\sin\theta_n & \cos\theta_n \end{bmatrix} \quad (n = 2, 4, 6) \tag{24}$$

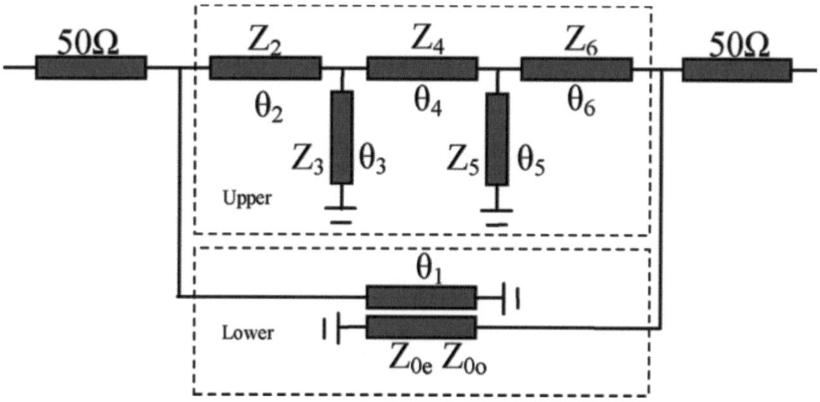

Figure 17.
Configuration of proposed UWB bandpass filter with shorted coupled lines.

$$M_m = \begin{bmatrix} 1 & 0 \\ \dfrac{1}{jZ_m \tan\theta_m} & 1 \end{bmatrix} (m = 3, 5) \tag{25}$$

$$Y_{lower} = -j\frac{1}{qZ_0}\begin{bmatrix} \cot\theta_1 & \dfrac{k}{\sin\theta_1} \\ \dfrac{k}{\sin\theta_1} & \cot\theta_1 \end{bmatrix} \tag{26}$$

where

$$k = \frac{(Z_{0e} - Z_{0o})}{(Z_{0e} + Z_{0o})}\ Z_0 = \sqrt{Z_{0o}Z_{0e}} \tag{27}$$

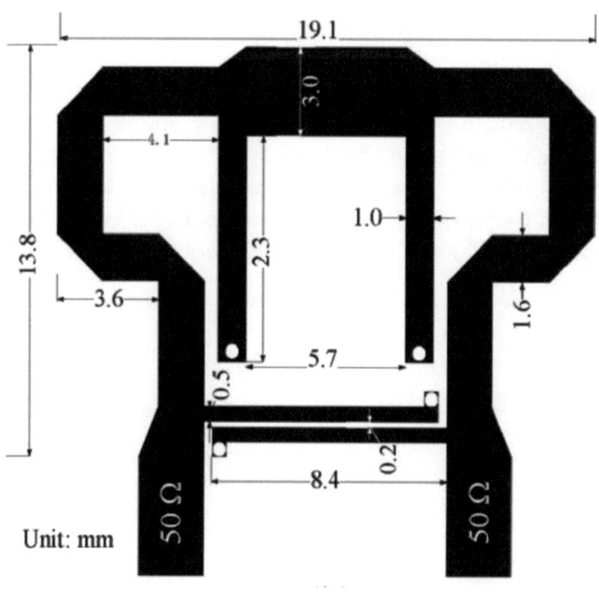

Figure 18.
Final circuit layout with dimension parameters of presented filter.

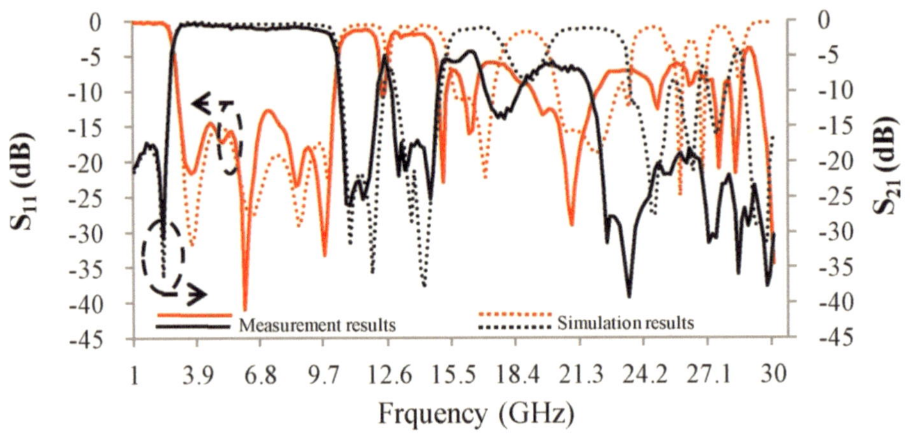

Figure 19.
Simulation and measurement results of presented UWB bandpass filter.

Thus, transmission coefficient can be derived. By properly changing the electrical lengths, UWB bandpass characteristics can be fulfilled, and the dimension parameters can be determined by using full-wave EM simulator, as demonstrated in **Figure 18**. Then simulation results and measurement results of fabricated UWB bandpass filter are shown in **Figure 19**, which shows excellent passband performance and multi-transmission zeros.

5. Microwave UWB bandpass filter with notch band

As illustrated in **Figure 2**, the UWB band has dramatic high FBW and surprisingly 7.5 GHz absolute bandwidth; therefore, the UWB has got great potential in many applications. However, a variety of undesired radio signal interferences and noise exist in the UWB frequency spectrum covering 3.1–10.6 GHz. Such as WiMAX (3.5 GHz), WLAN (5.2 GHz, 5.8 GHz), C-band satellite signals (5.975–6.745 GHz, 6.725–7.025 GHz), and RFID (6.8 GHz). As shown in **Figure 2**, some interference is introduced to the UWB communication system due to these narrowband signals. Thus, several notches are required to filter out the unwanted radio interference signals in UWB communication systems. In general, the methods of introducing a notch band in the UWB bandpass filter is of same essence, which is the electromagnetic energy of a certain frequency absorbed in the UWB band, so that signals with this frequency has been shorted out and averted to transmit from input to the output port. The design of the UWB notch filter has the following two

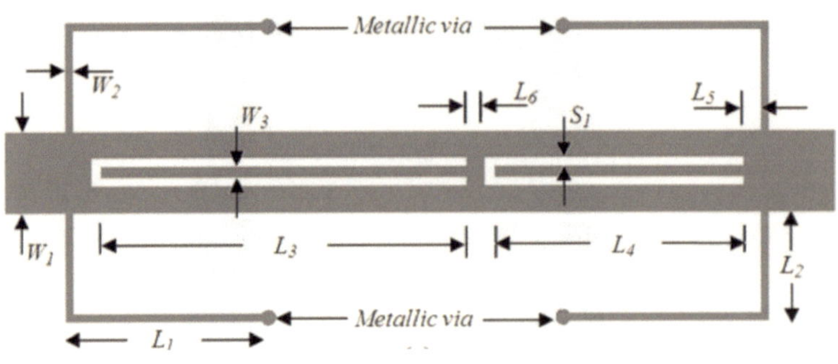

Figure 20.
Top view of the UWB bandpass filter with notch band.

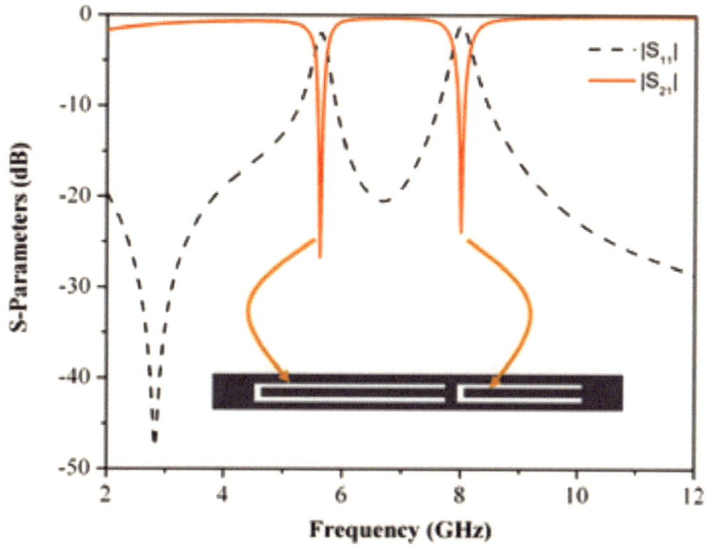

Figure 21.
The layout diagram of embedded open-circuited stubs and dual-notch band of proposed UWB filter.

methods: first, introducing additional notch unit circuits and, second, introducing lateral signal interference. The notch unit can be realized by various configurations, which includes single-mode/multimode resonator, a defected ground structure resonator, a metamaterial resonator, etc. Ultimately, the purpose of introducing notch unit circuits is to construct an electromagnetic absorption that set a notch in the UWB. Obviously, the notch which is designed based on aforementioned method is independently controllable. Furthermore, the number of notches can be easily extended, such as dual-notch band UWB filter and triple-notch band UWB filter.

To approach UWB bandpass characteristics with notch band, open-ended stubs can be applied to generate electromagnetic absorption [70]. The physical configuration is shown in **Figure 20**. Triple pairs of dumbbell defected ground structure are

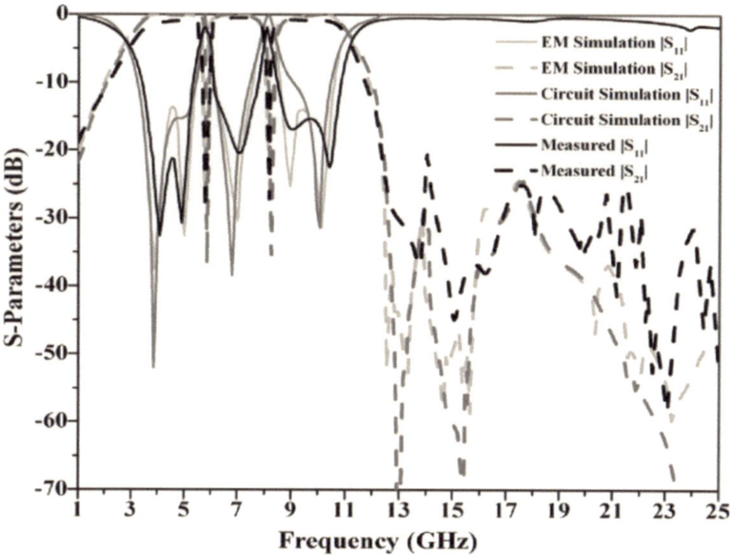

Figure 22.
EM simulation results versus circuit simulation results versus measurement results of presented UWB notch band bandpass filter.

introduced to realize low-pass transmission characteristics with improved out-off-band performance. Embedded open-circuited stubs are utilized to generate a pair of notch band, which is located at 5.75 and 8.05 GHz, as depicted in **Figure 21**. The developed filter is analyzed by using EM simulator CST microwave and fabricated on TACONIC substrate of dielectric constant 2.2. Excellent agreement can be observed which proves that the proposed UWB bandpass filter is of UWB with dual-notch band characteristics and wide stopband, as illustrated in **Figure 22**.

6. Conclusion

The research significance of the UWB bandpass filters and several conventional methods to achieve UWB bandpass filters with desired transmission performance is reviewed in this chapter. As the key building block of the UWB technology, the UWB bandpass filters can be realized by using several reliable design methodologies with excellent frequency response performance, which is of great value for scientific research and engineering significance.

Author details

Li-Tian Wang[1], Yang Xiong[2] and Ming He[1,3]*

1 College of Electronic and Optical Engineering, Nankai University, Tianjin, China

2 Southwest China Institute of Electronic Technology, Chengdu, China

3 China Tianjin Key Laboratory of Optoelectronic Sensor and Sensing Network Technology, Tianjin, China

*Address all correspondence to: heming@nankai.edu.cn

References

[1] Revision of Part 15 of the Commission's Rules Regarding Ultra-Wideband Transmission System. Washington, DC: FCC; 2002. ET-Docket 98-153

[2] Shannon CE. A mathematical theory of communication. Bell Labs Technical Journal. 1948;**27**(4):379-423. DOI: 10.1002/j.1538-7305.1948.tb00917

[3] Rappaport TS et al. Millimeter wave mobile communications for 5G cellular: It will work! IEEE Access. 2013;**1**: 335-349. DOI: 10.1109/ACCESS. 2013.2260813

[4] Gubbi J et al. Internet of Things (IoT): A vision, architectural elements, and future directions. Future Generation Computer Systems. 2013; **29**(7):1645-1660. DOI: 10.1016/j. future.2013.01.010

[5] Weng MH et al. Design of an ultra-wideband bandpass filter by using coupled three line microstrip structure. Journal of Electromagnetic Waves and Applications. 2012;**26**(5-6):716-728. DOI: 10.1080/09205071.2012.710798

[6] Weng MH et al. A high selectivity and wide stopband UWB bandpass filter using asymmetric SIRs with split-end. Microwave and Optical Technology Letters. 2014;**56**(6):1353-1356. DOI: 10.1002/mop.28321

[7] Wyndrum RW. Microwave filters, impedance-matching networks, and coupling structures. Proceedings of the IEEE. 1965;**53**(7):766-766. DOI: 10.1109/ PROC.1965.4048

[8] Pozar DM. Microwave Engineering. In: Hobeken J editor. NJ, USA: Wiley; 2006. DOI: 10.1109/45.31588

[9] Levy R, Cohn SB. A history of microwave filter research, design, and development. IEEE Transactions on Microwave Theory and Techniques. 1984;**32**(9):1055-1067. DOI: 10.1109/ tmtt.1984.1132817

[10] Hong J-S, Lancaster MJ. Microstrip Filters for RF/Microwave Applications. Hobeken, NJ, USA: John Wiley & Sons, Inc; 2001. DOI: 10.1002/0471221619

[11] Hsu CL, Hsu FC, Kuo JK. Microstrip bandpass filters for ultra-wideband (UWB) wireless communications. In: IEEE MTT-S International Microwave Symposium Digest, 2005. IEEE; 2005. DOI:10.1109/MWSYM.2005.1516698

[12] Yang GM et al. Design of ultra-wide band (UWB) bandpass filter based on defected ground structure. Microwave and Optical Technology Letters. 2010; **49**(6):1374-1377. DOI: 10.1002/ mop.22471

[13] Wu CH et al. A compact LTCC ultra-wideband bandpass filter using semi-lumped parallel-resonance circuits for spurious suppression. In: Microwave Conference, 2007. European IEEE; 2007. DOI: 10.1109/ EUMC.2007.4405245

[14] Han L, Wu K, Zhang X. Development of packaged ultra-wideband bandpass filters. IEEE Transactions on Microwave Theory and Techniques. 2010;**58**(1):220-228. DOI: 10.1109/tmtt.2009.2036399

[15] Qing-Xin C, Sai-Wai W. Ultra-wideband bandpass filter technologies. Journal of Microwaves. 2013;**29**:172-180

[16] Ting Z et al. A multi-mode resonator-based UWB bandpass filter with wide stopband. International Journal of Microwave and Wireless Technologies. 2016;**8**(7):5. DOI: 10.1017/S1759078715001026

[17] Ishii H et al. Development of UWB HTS bandpass filters with microstrip

stubs-loaded three-mode resonator. IEEE Transactions on Applied Superconductivity. 2013;**23**(3): 1500204-1500204. DOI: 10.1109/TASC.2012.2232697

[18] Lim TB, Zhu L. Highly selective differential-mode wideband bandpass filter for UWB application. IEEE Microwave and Wireless Components Letters. 2011;**21**(3):133-135. DOI: 10.1109/LMWC.2011.2104357

[19] Wu XH et al. Quintuple-mode UWB bandpass filter with sharp roll-off and super-wide upper stopband. IEEE Microwave and Wireless Components Letters. 2011;**21**(12):661-663. DOI: 10.1109/LMWC.2011.2170672

[20] Taibi A et al. A novel design method for compact UWB bandpass filters. IEEE Microwave and Wireless Components Letters. 2015;**25**(1):4-6. DOI: 10.1109/LMWC.2014.2363016

[21] Lan SW et al. Design of a compact ultra-wideband bandpass filter with an extremely broad stopband region. IEEE Microwave and Wireless Components Letters. 2016;**26**(6):1-3. DOI: 10.1109/LMWC.2016.2558039

[22] Gao X, Feng W, Che W. Compact ultra-wideband bandpass filter with improved upper stopband using open/shorted stubs. IEEE Microwave and Wireless Components Letters. 2017;**27**(2):123-125. DOI: 10.1109/LMWC.2016.2647385

[23] Tianliang Z et al. High-temperature superconducting multimode ring resonator ultrawideband bandpass filter. IEEE Microwave and Wireless Components Letters. 2018:1-3. DOI: 10.1109/LMWC.2018.2845116

[24] Yao B et al. Compact UWB bandpass filter with improved upper-stopband performance using multiple-mode resonator. IEEE Microwave and Wireless Components Letters. 2009;**19**(1):27-29. DOI: 10.1109/LMWC.2008.2008558

[25] Zhang Z, Xiao F. An UWB bandpass filter based on a novel type of multi-mode resonator. IEEE Microwave and Wireless Components Letters. 2012;**22**: 506-508. DOI: 10.1109/LMWC.2012.2218229

[26] Hong J-S, Shaman H. An optimum ultra-wideband microstrip filter. Microwave and Optical Technology Letters. 2010;**47**(3):230-233. DOI: 10.1002/mop.21133

[27] Wong WT et al. Highly selective microstrip bandpass filters for ultra-wideband (UWB) applications. In: Asia-Pacific Microwave Conference Proceedings, 2005 (APMC 2005). IEEE; 2006

[28] Shaman H, Hong JS. A novel ultra-wideband (UWB) bandpass filter (BPF) with pairs of transmission zeroes. IEEE Microwave and Wireless Components Letters. 2007;**17**(2):121-123. DOI: 10.1109/LMWC.2006.890335

[29] Shaman H, Hong JS. An optimum ultra-wideband (UWB) bandpass filter with spurious response suppression. In: IEEE Wireless & Microwave Technology Conference; 2007. DOI: 10.1109/WAMICON.2006.351902

[30] Deng HW et al. Compact quintuple-mode stub-loaded resonator and UWB filter. IEEE Microwave and Wireless Components Letters. 2010;**20**(8): 438-440. DOI: 10.1109/LMWC.2010.2049481

[31] Zhu H, Chu QX. Compact ultra-wideband (UWB) bandpass filter using dual-stub-loaded resonator (DSLR). IEEE Microwave and Wireless Components Letters. 2013;**23**(10): 527-529. DOI: 10.1109/LMWC.2013.2278278

[32] Chu QX, Wu XH, Tian XK. Novel UWB bandpass filter using stub-loaded

multiple-mode resonator. IEEE Microwave and Wireless Components Letters. 2011;**21**(8):403-405. DOI: 10.1109/lmwc.2011.2160526

[33] Zhu L, Sun S, Menzel W. Ultra-wideband (UWB) bandpass filters using multiple-mode resonator. IEEE Microwave and Wireless Components Letters. 2005;**15**(11):796-798. DOI: 10.1109/lmwc.2005.859011

[34] Wong SW, Zhu L. Quadruple-mode UWB bandpass filter with improved out-of-band rejection. IEEE Microwave and Wireless Components Letters. 2009;**19**(3):152-154. DOI: 10.1109/LMWC.2009.2013735

[35] Chu QX, Tian XK. Design of UWB bandpass filter using stepped-impedance stub-loaded resonator. IEEE Microwave and Wireless Components Letters. 2010;**20**(9):501-503. DOI: 10.1109/LMWC.2010.2053024

[36] Lu X et al. Superconducting ultra-wideband (UWB) bandpass filter design based on quintuple/quadruple/ triple-mode resonator. IEEE Transactions on Microwave Theory and Techniques. 2015;**63**(4):1281-1293. DOI: 10.1109/tmtt.2015.2402152

[37] Shang Z et al. Design of a superconducting ultra-wideband (UWB) bandpass filter with sharp rejection skirts and miniaturized size. IEEE Microwave and Wireless Components Letters. 2013;**23**(2):72-74. DOI: 10.1109/LMWC.2013.2239633

[38] Ishida H, Araki K. Design and analysis of UWB band pass filter with ring filter. In: 2004 IEEE MTT-S International IEEE Microwave Symposium Digest; 2004. DOI: 10.1109/MWSYM.2004.1338806

[39] Sun S, Zhu L. Wideband microstrip ring resonator bandpass filters under multiple resonances. IEEE Transactions on Microwave Theory and Techniques. 2007;**55**(10):2176-2182. DOI: 10.1109/tmtt.2007.906510

[40] Lok U, Chiou YC, Kuo JT. Quadruple-mode coupled-ring resonator bandpass filter with quasi-elliptic function passband. IEEE Microwave and Wireless Components Letters. 2008;**18**(3):179-181. DOI: 10.1109/LMWC.2008.916782

[41] Wong SW, Zhu L, Quek LC, Chen ZN. A stopband-enhanced UWB bandpass filter using short-/open-stubs embedded ring resonator. In: IEEE Microwave Conference; 2009. DOI: 10.1109/APMC.2009.5384314

[42] Kim CH, Chang K. Ring resonator bandpass filter with switchable bandwidth using stepped-impedance stubs. IEEE Transactions on Microwave Theory and Techniques. 2010;**58**(12): 5609230. DOI: 10.1109/tmtt.2010.2084584

[43] Fan J et al. Wideband microstrip bandpass filter based on quadruple mode ring resonator. IEEE Microwave and Wireless Components Letters. 2012; **22**(7):348-350. DOI: 10.1109/LMWC.2012.2199977

[44] Kim CH, Chang K. Ultra-wideband (UWB) ring resonator bandpass filter with a notched band. IEEE Microwave and Wireless Components Letters. 2011; **21**(4):206-208. DOI: 10.1109/LMWC.2011.2109942

[45] Xu KD et al. Compact ultra-wideband bandpass filter using quad-T-stub-loaded ring structure. Microwave and Optical Technology Letters. 2014;**56**(9):1988-1991. DOI: 10.1002/mop.28508

[46] Oh S, Song J, Lee J. UWB bandpass filter based on ring resonator. Microwave and Optical Technology Letters. 2013;**55**(9):2047-2051. DOI: 10.1002/mop.27780

[47] Li X, Ji X. Novel compact UWB bandpass filters design with cross-coupling between Lambda/4 short-circuited stubs. IEEE Microwave and Wireless Components Letters. 2014; **24**(1):23-25. DOI: 10.1109/LMWC.2013.2287231

[48] Abbosh AM. Design method for ultra-wideband bandpass filter with wide stopband using parallel-coupled microstrip lines. IEEE Transactions on Microwave Theory and Techniques. 2012;**60**(1):31-38. DOI: 10.1109/tmtt.2011.2175241

[49] Luo X et al. Compact ultra-wideband (UWB) bandpass filter with ultra-narrow dual- and quad-notched bands. IEEE Transactions on Microwave Theory and Techniques. 2011;**59**(6): 1509-1519. DOI: 10.1109/tmtt.2011.2116800

[50] Nedil M et al. Ultra-wideband bandpass filters using multilayer slot coupled transitions. Journal of Electromagnetic Waves and Applications. 2008;**22**(4):501-516. DOI: 10.1163/156939308784150353

[51] Hao ZC, Hong JS. Ultra wideband bandpass filter using embedded stepped impedance resonators on multilayer liquid crystal polymer substrate. IEEE Microwave and Wireless Components Letters. 2008;**18**(9):581-583. DOI: 10.1109/LMWC.2008.2002448

[52] Hao ZC, Hong JS. Compact ultra-wideband bandpass filter using broadside coupled hairpin structures on multilayer liquid crystal polymer substrate. Electronics Letters. 2008; **44**(20):1197-1198. DOI: 10.1049/el:20089877

[53] Hao ZC, Hong JS. Quasi-elliptic UWB bandpass filter using multilayer liquid crystal polymer technology. IEEE Microwave and Wireless Components Letters. 2010;**20**(4):202-204. DOI: 10.1109/LMWC.2010.2042551

[54] Hao ZC, Hong JS. UWB bandpass filter using cascaded miniature high-pass and low-pass filters with multilayer liquid crystal polymer technology. IEEE Transactions on Microwave Theory and Techniques. 2010;**58**(4):941-948. DOI: 10.1109/tmtt.2010.2042632

[55] Li Y et al. Novel multilayered ultra-broadband bandpass filters on high-impedance slotline resonators. IEEE Transactions on Microwave Theory and Techniques. 2019;**67**(1):129-139. DOI: 10.1109/TMTT.2018.2873330

[56] Honari MM et al. Two-layered substrate integrated waveguide filter for UWB applications. IEEE Microwave and Wireless Components Letters. 2017;**27** (7):1-3. DOI: 10.1109/lmwc.2017.2711510

[57] Hao ZC, Hong JS. Ultra-wideband bandpass filter using multilayer liquid-crystal-polymer technology. IEEE Transactions on Microwave Theory and Techniques. 2008;**56**(9):2095-2100. DOI: 10.1109/tmtt.2008.2002228

[58] Xu J et al. Compact UWB bandpass filter with a notched band using radial stub loaded resonator. IEEE Microwave and Wireless Components Letters. 2012; **22**(7):351-353. DOI: 10.1109/LMWC.2012.2201930

[59] Gomez-Garcia R et al. High-order planar bandpass filters with electronically-reconfigurable passband width and flatness based on adaptive multi-resonator cascades. IEEE Access. 2019;**99**:1-1. DOI: 10.1109/ACCESS.2019.2891889

[60] Ghatak R et al. A compact UWB bandpass filter with embedded sir as band notch structure. IEEE Microwave and Wireless Components Letters. 2011; **21**(5):261-263. DOI: 10.1109/LMWC.2011.2128302

[61] Xu HX, Wang GM, Zhang CX. Fractal-shaped UWB bandpass filter based on composite right/left handed

transmission line. Electronics Letters. 2010;**46**(4):285-287. DOI: 10.1049/el.2010.3139

[62] Wei F et al. Compact UWB bandpass filter with dual notched bands based on SCRLH resonator. IEEE Microwave and Wireless Components Letters. 2011;**21**(1):28-30. DOI: 10.1109/LMWC.2010.2088113

[63] Shaman H. Ultra-wideband (UWB) bandpass filter with embedded band notch structures. IEEE Microwave and Wireless Components Letters. 2007;**17**: 193-195. DOI: 10.1109/LMWC.2006.890467

[64] Yang GM et al. Small ultra-wideband (UWB) bandpass filter with notched band. IEEE Microwave and Wireless Components Letters. 2008; **18**(3):176-178. DOI: 10.1109/LMWC.2008.916781

[65] Wang H et al. Cross-shaped UWB bandpass filter with sharp skirt and notched band. Electronics Letters. 2012; **48**(2):96-97. DOI: 10.1049/el.2011.3694

[66] Hao Z-C et al. Ultra-wideband bandpass filter with multiple notch-bands on multilayer liquid crystal polymer substrate. IET Microwaves, Antennas and Propagation. 2009;**3**(5):749-756. DOI: 10.1049/iet-map.2008.0232

[67] Luo X et al. Compact UWB bandpass filter with ultra narrow notched band. IEEE Microwave and Wireless Components Letters. 2010;**20**(3):145-147. DOI: 10.1109/LMWC.2010.2040212; 10.1109/tmtt.2011.2116800

[68] Gupta SC, Kumar M, Meena RS. Design & analysis of a microstrip line multi band UWB filter. AEU - International Journal of Electronics and Communications. 2016;**70**:1556-1564. DOI: 10.1016/j.aeue.2016.09.014

[69] Yan T et al. High-selectivity UWB bandpass filter with a notched band using stub-loaded multi-mode resonator. AEU - International Journal of Electronics and Communications. 2016;**70**(12):1617-1621. DOI: 10.1016/j.aeue.2016.09.016

[70] Sarkar P et al. Compact UWB bandpass filter with dual notch bands using open circuited stubs. IEEE Microwave and Wireless Components Letters. 2012;**22**(9):453-455. DOI: 10.1109/LMWC.2012.2210395

[71] Pirani S, Nourinia J, Ghobadi C. Band-notched UWB BPF design using parasitic coupled line. IEEE Microwave and Wireless Components Letters. 2010;**20**(8):444-446. DOI: 10.1109/LMWC.2010.2049830

[72] Zhang XY, Zhang YW, Xue Q. Compact band-notched UWB filter using parallel resonators with a dielectric overlay. IEEE Microwave and Wireless Components Letters. 2013; **23**(5):252-254. DOI: 10.1109/LMWC.2013.2255121

[73] Song K et al. Compact ultra-wideband (UWB) bandpass filters with multiple notched bands. IEEE Microwave and Wireless Components Letters. 2010;**20**(8):447-449. DOI: 10.1109/LMWC.2010.2050303

[74] Wei F et al. Compact UWB bandpass filter with triple-notched bands using triple-mode stepped impedance resonator. IEEE Microwave and Wireless Components Letters. 2012; **22**(10):512-514. DOI: 10.1109/LMWC.2012.2215845

[75] Zhao J et al. Compact microstrip UWB bandpass filter with dual notched bands using e-shaped resonator. IEEE Microwave and Wireless Components Letters. 2013;**23**(12):638-640. DOI: 10.1109/LMWC.2013.2283873

[76] Sekar V, Entesari K. Miniaturized UWB bandpass filters with notch using slow-wave CPW multiple-mode

resonators. IEEE Microwave and Wireless Components Letters. 2011; **21**(2):80-82. DOI: 10.1109/LMWC.2010.2096803

[77] Song Y, Yang GM, Geyi W. Compact UWB bandpass filter with dual notched bands using defected ground structures. IEEE Microwave and Wireless Components Letters. 2014; **24**(4):230-232. DOI: 10.1109/LMWC.2013.2296291

Chapter 6

Passive Components for Ultra-Wide Band (UWB) Applications

Dalia N. Elsheakh and Esmat A. Abdallah

Abstract

UWB technology brings the convenience and mobility of wireless communications to very high-speed interconnects in the home and office due to the precision capabilities combined with the low power. This makes it ideal for certain radio frequency sensitive environments such as hospitals and healthcare as well as radars. UWB intrusion-detection radar is used for detecting through the wall and also used for security with fuse avoidance radar, precision locating and tracking (using distance measurements between radios), and precision time-of-arrival-based localization approaches. The FCC issued a ruling in 2002 that allowed intentional UWB emissions in the frequency range between 3.1 and 10.6 GHz, subject to certain restrictions for the emission power spectrum. Other definitions for ultra-wideband range of frequency are also used such as any device that has 500 MHz bandwidth or fractional bandwidth greater than 25% is considered an UWB enable high data rate to be transferred with a very low power that does not exceed −41.3 dBm.

The main advantage of UWB technology is that it has the ability to transmit high bandwidth data between various devices with distances in the order of 10 m far from each other, such as home or office appliances with high sped transfer which may reach 1Gbit/s.

There are many components that are designed and used in UWB systems such as antennas, power dividers/combiners, filters (LPF, BPF, etc.), rectennas, filtennas, etc.

Many types of antennas are able to achieve UWB. Monopole antennas are usually used as linearly polarized antennas which prove to be the best whelming choice for use in various automobiles and mobile equipment. Log parodic and Yagi antennas are other types of UWB with high gain. Electromagnetic band gap (EBG) structure as defected ground or split ring resonator are also used to improve the antenna bandwidth and achieve UWB.

Many UWB filters are designed and implemented using modern techniques such as meta-material, tuning stubs, defected ground structures, modified CMRC (compact microstrip resonant cell), etc. These filters may be switchable or tunable in order to make notch frequencies within the passband to provide interference immunity from unwanted radio signals, such as wireless local area networks (WLAN) and worldwide interoperability for microwave access (WiMAX) that cohabit within the UWB spectrum.

The need of compact multiband filtenna (filter combined with antenna) with the ability of covering the current standards at the microwave band and the next generation standards at the millimeter wave band simultaneously is raising. Also, the operation of cognitive radio and self-adaptive systems need to dynamically monitor the frequency spectrum in search of the unused licensed channels. All these applications need different types of filtenna which may be fixed, switchable or tunable that can operate in the

IntechOpen

UWB range of frequency and also with narrow band modes to cover for example UWB/WiMAX applications.

Keywords: ultra-wide band (UWB), monopole antenna, UWB filters, filtenna, electromagnetic bandgap structure (EBG), log parodic

1. Introduction

In recent years, UWB technology has mostly focused on consumer electronics and wireless communications. Federal Communication Commission (FCC) issued a report in February 2002, allowing the commercial and unlicensed deployment of UWB applications in USA for both indoor and outdoor spectral mask. This wide frequency allocation initiated a lot of researches from both industry and academia [1]. UWB is used for short and medium range of radio communications and positioning applications.

The European regulatory body issue similar restrictions are shown in **Figure 1**. The key limitations for wireless communication using UWB are mentioned in [2–4].

UWB impulse radio system has several advantages over other conventional systems [5].

i. High data rate wireless transmission: UWB systems can support more than 500 Mb/s data transmission rate in the range of 10 m, which enables for new services and applications.

ii. High precision ranging: UWB systems have good time-domain resolution and it could be provided centimeter accuracy for location and tracking applications.

iii. UWB is used for location and tracking applications with cm accuracy.

iv. UWB can operate under LOS and NLOS environments for signal penetrating obstacles.

v. UWB system is capable of resistance to multipath fading.

vi. The power spectral density is very low so it is secure and can coexist with other services such as WLAN, GPS, cellular system, etc.

vii. The UWB system has low cost due to using CMOS technology.

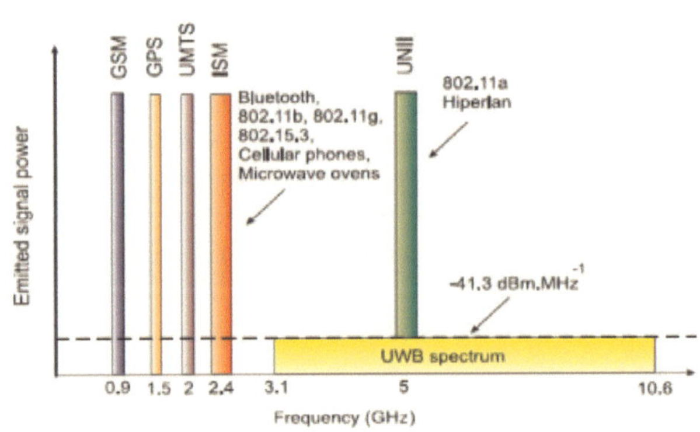

Figure 1.
European regulatory body spectrum.

Due to UWB technology and using nanosecond pulses in many applications such as military and biotechnology applications [6–8], the need for very broadband circularly polarized antenna has emerged. These UWB CP antennas are the substitution of the narrowband CP microstrip patch antennas [8–10].

One of the most commonly used devices to control the spectrum of radio frequency signals and necessary in an UWB radio system whether in impulse system or multiband system, in order to reconfigure the UWB signal to satisfy the spectrum regulation is the filter. Ultra-wide band (UWB) band-pass filter that achieves ultra bandwidth from 3.1 to 10.6, low insertion loss, low and flat group delay, out-band performance can be considered a well-designed UWB band-pass filter.

In UWB band-pass filters, one can use many techniques in their design such as composite low-pass and high-pass structure [11], multiple-mode resonator structure [14], and short-circuited stub [12, 13]. Because UBW components occupy a large bandwidth which may be extended from 3.1 up to 10.6 GHz, interference attenuation due to coexisting services should be avoided. This is the motivation of using switchable or tunable narrow band notch within the passband of the UWB filter [15–17]. This may be achieved by many methods such as using additional notch resonators [18], embedded open stubs [19], asymmetric coupled fed lines [20], out-of-phase transmission cancelation [21], meander-line slots [22], and short-circuited stub resonators in a multilayer periodical structure [23].

To deal with different co-existed communication needs, the reconfigurable notch band implementation is required, but little research is concentrated on a UWB BPF with reconfigurable, switchable, or tunable notch bands as in [24–26].

An HFSS, FEM-based, 3D full wave electromagnetic solver simulator by ANSYS as well as CST were used for the design of all designed antennas. Also, ADS was used to design the filters and filtennas in addition to the above electromagnetic simulators in this chapter.

This chapter describes one example for printed millimeter wave antenna implementations, illustrating specific and interesting particular solutions for their design and two shapes of single UWB antenna in radio frequency range. In addition, two examples of UWB filters and one example of UWB filtennas are introduced.

2. UWB millimeter wave antenna

Classical antenna as reflector, lens, and horn type antennas have been used in millimeter-wave devices. But for low-cost, these antennas are commercial expensive devices, bulky, heavy and require complex feeding in an array system. In addition, they are very difficult to integrate with solid-state devices [6–9]. However, the microstrip antennas (MPAs) are narrow bandwidth and are large size about half-wavelength structures.

2.1 UWB slot antennas for linear and circular polarizations

Three different types of broad multi-band linearly and circularly polarized slot antennas (rectangular-, circular-, and triangular-shaped slots) for millimeter wave wireless communication applications [27] are shown in **Figures 2** and **3**, respectively. Proposed antenna consists of a slot radiator on the top metal layer and coupled to a rectangular dielectric resonator above the slot. The conventional microstrip-line-feed is used for different shapes of slot antennas. Final designed antennas were fabricated, and their characteristics were measured as reflection coefficient. The bandwidth of $|S_{11}| < -10$ dB was extended from 19.5 up to 75 GHz. This band covers wireless MM-wave applications and wireless networks, and the

WLAN, WPAN, and W-bands and most of 5th Generation mobile [28–32]. The average radiation efficiency and gain over the entire operating band are about 60% and 6 dBi, respectively [27]. Printed different shapes of slot antenna show small dimensions ($L_g \times W_g$) cut at different shapes of slots on larger conductor and are centered above the microstrip-feed line. The microstrip feed line is composed of a straight section of length L_f. To improve the antenna reflection coefficient response, a square stub slot is added with the side length S = 0.6 mm for further improvement in the antenna impedance matching as shown in **Figure 2**. The width of the tuning line is equal to that of the 50 Ω microstrip line (W_f = 0.56 mm). The optimized dimensions of the proposed antennas are shown in **Table 1**.

2.1.1 Linearly polarized antennas

The first shape of antenna design is either square with side length = 3.5 mm or rectangular slot with dimensions L = 2.9 mm and W = 4 mm as shown in **Figure 2 (a)**. The antenna reflection coefficient $|S_{11}|$ is shown in **Figure 4(a)**. **Figure 4(a)** shows that two antenna designs were examined with the same start and end of the operating resonate frequency band from 19.5 to 75 GHz. The square slot shaped antenna has bandwidth discontinuities from 34 to 45 GHz. The second shape included in this study is the circular and elliptical shaped slot as shown in

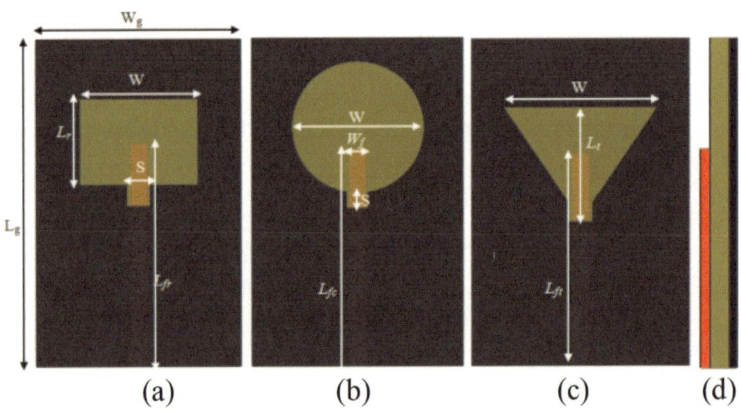

Figure 2.
Different shapes of slot antenna for LP (a) rectangular, (b) circular, (c) triangular, and (d) side view [27].

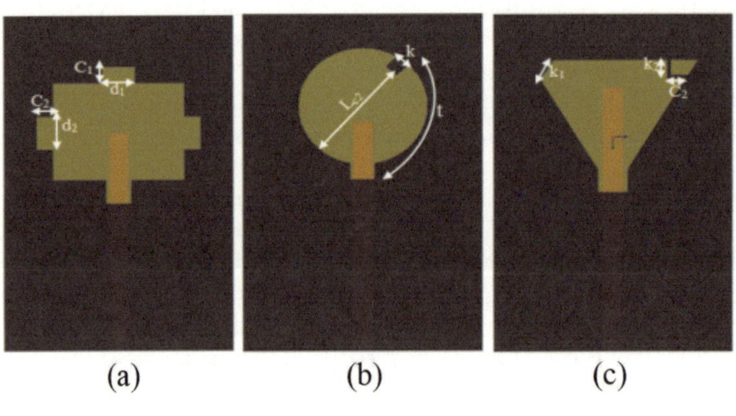

Figure 3.
Shapes of slot antenna for circular polarization (a) rectangular, (b) circular, and (c) triangular [27].

W_g	L_g	W	L_r	L_{ftc}	K	L_t	K_2	d_1	t
7	10	4	2.9	11.6	0.25	4.5	0.6	0.7	4.5
L_{fr}	L_{fc}	W_f	S	L_t	C_1	C_2	K_2	d_1	t
8.6	8.5	0.56	0.6	4	0.7	0.7	0.6	0.7	4.5

Table 1.
The optimized antennas (all dimensions in mm) [27].

Figure 2(b). The ellipse slot major diameter is W = 4 mm, and different radius ratios were used including the design with a ratio equal to 1.35 as shown in **Figure 4(b)**. The $|S_{11}|$ response shows that the optimal bandwidth at $|S_{11}| < -10$ dB is obtained using a circular slot shape. The bandwidth extends from 21 to 80 GHz in this case, while for the case of the elliptical slot, the bandwidth starts from 28 GHz and has the same end at 80 GHz as shown in **Figure 4(b)**. The triangular shape is the third slot shape used in this study, and it is compared to trapezoidal as an intermediate stage between a triangle and a rectangle slot as shown in **Figure 2(c)**. **Figure 4(c)** shows that both of the stated shapes resonate at frequencies higher than the other previous shapes, as it starts from 29 GHz.

2.1.2 Slot antennas for circularly polarized

Single-feed MPAs for CP are usually achieved by using traditional simple changes in the shape of the patch such as truncating corners, using nearly square or nearly circular patches, cutting a diagonal slot in the square or circular patches, protruding or inserting a pair of symmetric perturbation elements at the boundary of a circular patch [33–35]. However, this type produced narrow axial ratio (AR) bandwidth. The dual-fed and sequential array structure produced wider AR bandwidth, but this requires more complicated design and may occupy larger space. Slot antennas are expected to overcome the limited bandwidth problem as well as similar changes in the slot shape could be used to achieve CP.

Rectangular notches etched in the rectangular slot are used as a way to improve the 3 dB AR bandwidth as shown in **Figure 5(a)**. The notch width $C_1 = C_2 = 0.7$ mm gives the best wideband AR. The comparison of AR between measured and simulated is also shown in **Figure 5(a)**.

In the proposed design, a square stub with dimension side K was added to the circular shaped slot to achieve the CP performance. The stub is added at a radial distance t from square stub. The locations of this stub set the AR bandwidth without degrading the antenna performance as shown in **Figure 5(b)**. This figure shows that the stub with notch at a distance t = 3 mm gives the best performance AR

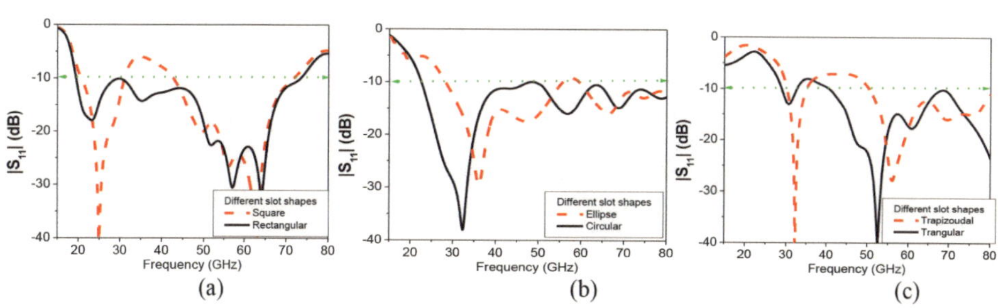

Figure 4.
Antenna $|S_{11}|$ response with different slot shapes (a) rectangular, (b) circular, and (c) triangular [27].

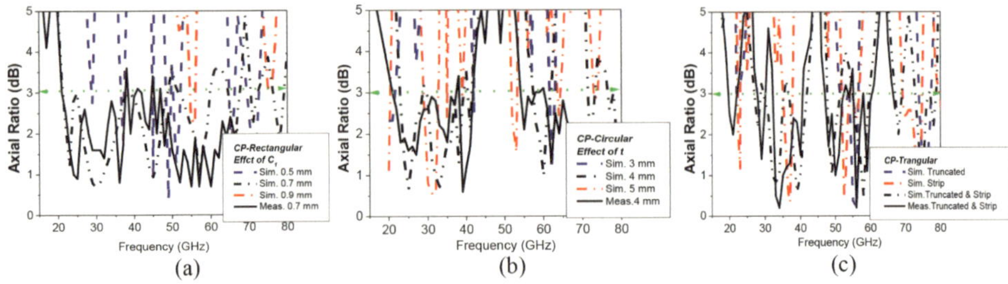

Figure 5.
Axial ratio of the antennas with different slot shapes (a) rectangular, (b) circular, and (c) triangular [27].

bandwidth. However, AR bandwidth discontinuities appear from 40 to 55 GHz. The comparison results of AR values simulated and measured at t = 3 mm is shown in **Figure 5(b)**. Finally, for the triangular slot shape, a truncated corner was used. To improve the AR bandwidth, an L-shaped strip was added at the other triangular corner with width 0.1 mm as shown in **Figure 3**. The bandwidth for simulated and measured AR values for the triangular CP antenna is shown in **Figure 5(c)**. From previous shapes, it appears that rectangular shaped slot with notches gives wide axial ratio bandwidth without degrading the antenna bandwidth.

2.1.3 Experimental results and discussion

The proposed antenna designs were fabricated by using milling machine technology with 0.1 mm accuracy on Rogers 6035HTC substrate with a 0.25 dielectric thickness and 0.017-mm copper thickness. A 1.85-mm end launcher connector is used to measure the proposed antennas. The simulation reflection coefficient was verified by comparison with the experimental results of the antennas by using 37397C Anritsu vector network analyzer. Photos of the fabricated antennas are shown in **Figures 6** and 7. The comparison between measured and simulated $|S_{11}|$ for linearly and circularly proposed antennas are shown in **Figure 8(i)** and **(ii)**, respectively. The measurement and simulation result data are in a good agreement. Measured results ended at 65 GHz as it is the end-point of the network analyzer. The rectangular slot shape gives the best antenna performance for linearly and circularly polarized slots. These designs also have low profile, wide impedance bandwidth $|S_{11}| < -10$ dB, and wide 3 dB axial ratio.

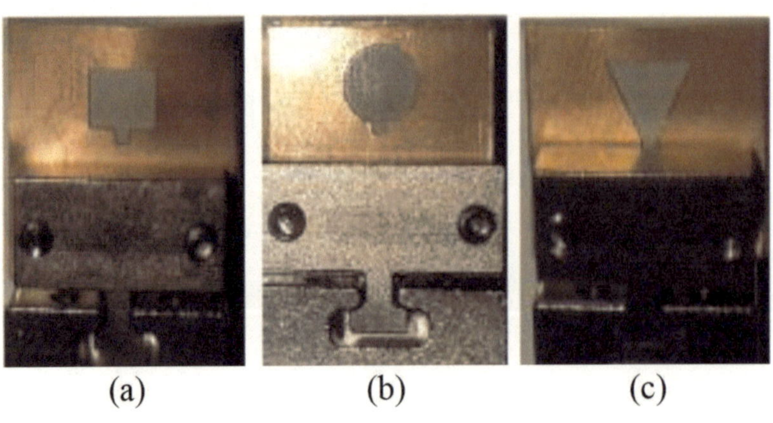

Figure 6.
Photo of fabricated linear polarized slot antenna, (a) rectangular, (b) circular, and (c) triangular shaped [27].

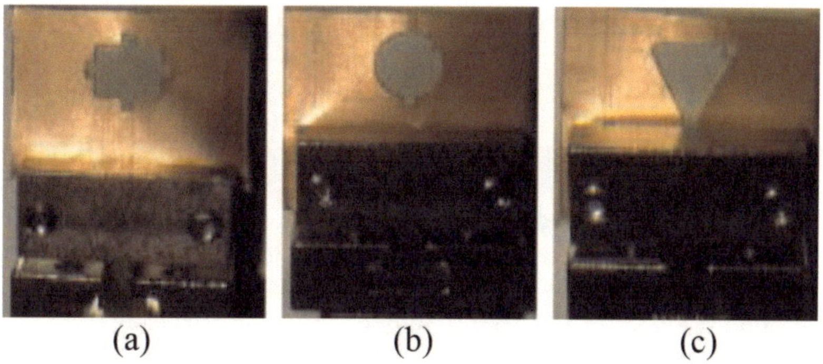

Figure 7.
Photo of fabricated circular polarized slot antenna, (a) rectangular, (b) circular, and (c) triangular shaped [27].

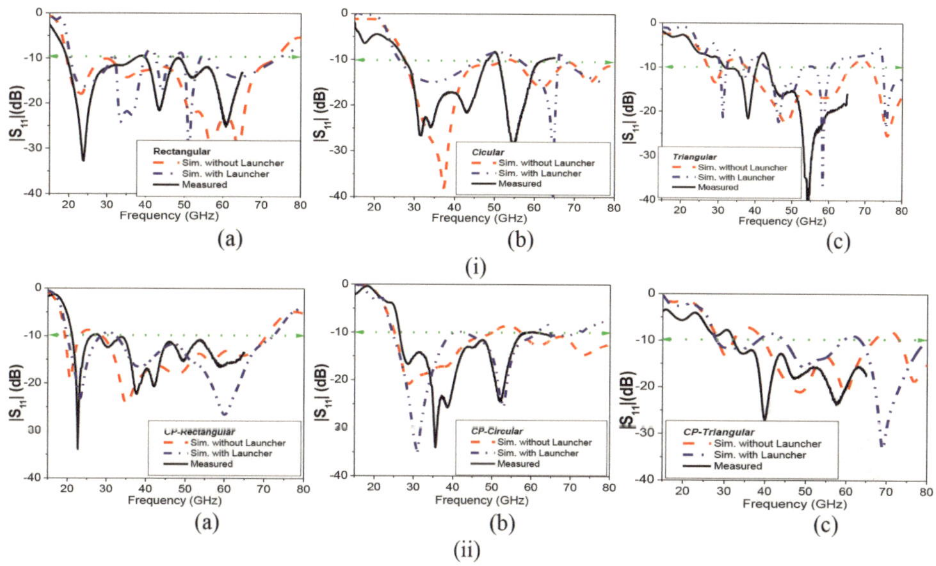

Figure 8.
Measured and simulated $|S_{11}|$ of (i) linear polarized (a) rectangular, (b) circular, and (c) triangular slot shaped and (ii) circular polarized (a) rectangular, (b) circular, and (c) triangular slot shaped [27].

3. UWB antenna in radio frequency range

Planar microwave circuitry has generated attractive radiating structures with high gain, low weight, reliability, ease of manufacturing and integration such as the Vivaldi antennas [36, 37], and the tapered slot antenna [38] for UWB antennas. The printed planar log-periodic dipole (LPDA) is the most suitable solution microwave frequencies [39]. LPDAs have a lot of advantages, such as directive radiation pattern, linear polarization and low cross polarization ratio over a wide frequency range [5]. At the beginning, coaxial cable was used for feeding the printed LPDAs at the radio and the TV frequency bands; however, it was found that the performance became worse when frequency increases. Due to huge increase in data traffic, there is a requirement for wireless networks which support both data and voice transfer simultaneously for short-range wireless communication systems [1, 2].

3.1 UWB CPW-fed LPDA for wireless applications

This section presents a LPDA for UWB applications [39]. The antenna consists of cascading four U-shaped elements of different line lengths with balun circuit to improve the antenna impedance matching. The proposed antenna area dimensions are 50×50 mm^2 built on FR4 substrate thickness 0.8 mm. Simulator HFSS is used for modeling the designed antenna. The pulse distortion is verified by measuring the proposed antenna performance with virtually steady group delay. The simulation and measurement results show that the designed antenna exhibits good impedance matching, stable radiation patterns throughout the whole operating frequency bands, acceptable gain and stable group delay over the entire operating band.

LPDA is UWB with the multiple resonance property; its bandwidth can be enhanced by increasing the number of the dipole elements [40–42]. Balanced structure, CPW fed antennas are very good candidates since the feed lines and the slots are on the same side of the substrate. There are many researches done to design LPDA but most of the published papers are not compact and their length is near from wavelength. The compact antenna dimensions, as shown in **Figure 9**, are $50 \times 50 \times 0.8$ mm^3. The proposed USLPDA antenna introduces USUWB with the multiple resonant property and compact size compared to earlier designs where UWB was realized using a rectangular slot [43], U-shaped dipole elements or stubs [38]. The bandwidth of this antenna at −10 dB reflection coefficient extends from 1.85 to 11 GHz which is wide enough to cover the FCC approved UWB in addition to wireless communications. The antenna exhibits good performance and can operate at wireless applications.

3.1.1 Geometry antenna and design

The designed geometry antenna is shown in **Figure 9**; the antenna is composed of four different lengths with U-shaped stubs. The lengths and spacing of the elements of LPDA increase logarithmically from one end to the other. The design of the LPDA is used where a wide range of frequencies is needed while still having moderate gain and directionality. The simulator HFSS is used to validate and optimize by simulating the designed antenna. The designed antenna is built on a commercial FR4 substrate with dielectric constant $\varepsilon_r = 4.6$, and loss tangent $\tan \delta = 0.02$. The antenna is fed by a 50 Ω transmission line, which can be easily integrated with other microwave circuits printed on the same substrate. For designing procedure,

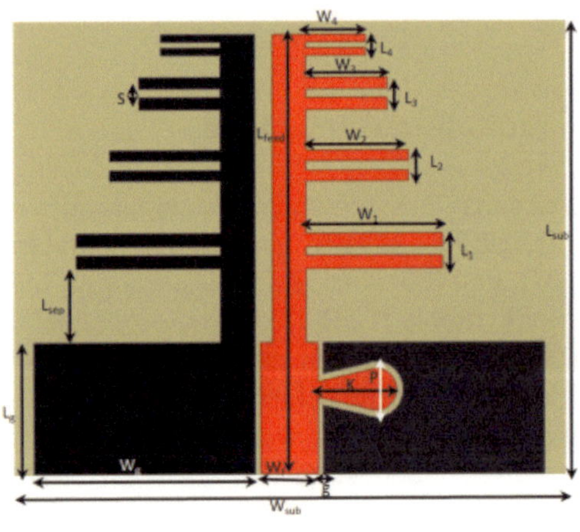

Figure 9.
Layout of the proposed log periodic dipole antenna (semi-LPDA) [39].

the number of trial steps is needed, the scale-factor τ, spacing factor δ, and the number of the dipole elements N should be determined. Second, the length of the longest arm, which responses to the lowest resonance frequency f_1, should be computed by following Eqs. (1)–(6) [39].

$$\frac{W_{i-1}}{W_i} = \tau \tag{1}$$

$$\delta = \frac{L_{isep}}{4W_i} \tag{2}$$

$$W_1 = \frac{\lambda_{1,eff}}{4} \tag{3}$$

$$\mathrm{N} = 1 - (\ln \mathrm{B_s}/\ln \tau) \tag{4}$$

$$\mathrm{B_a} = 1.1 + 30.7\,\delta\,(1 - \tau) \tag{5}$$

$$\mathrm{B_s} = \mathrm{B_a B_o} \tag{6}$$

where λ_{1eff}, B_o, N, and int i are the longest effective operating wavelength, the operating frequency, number of elements, and i is an integer that varies from 2 to 5, respectively. To improve the impedance, matching the balun circuit with suitable dimensions is used as shown in **Figure 9**. **Table 2** introduces the dimensions of the proposed antenna [39].

3.1.2 Measured results

The designed antenna is fabricated by using photolithographic technique, as shown in **Figure 10**, and parameter performance is measured. The simulated and measured input reflection coefficient of the antennas is in good agreement, as shown in **Figure 10(b)**. The designed dipole impedance bandwidth at −10 dB of antenna extended from 1.85 to 11 GHz to cover most of the wireless applications and FCC UWB regulation. The antenna gain data are compared between measured and simulated results as shown in **Figure 11**. The designed LPDA antenna achieves simulated average gain 5.5 dBi, and the peak realized gain around 6.5 dBi at 2.7 GHz as shown in **Figure 11(a)**. The measured results show very good agreement with simulated results and about ±3 dBi difference on average over the operating band. Wheeler cap method [44] can be used to calculate the antenna radiation efficiency. The average radiation efficiency is around 70% over the operating band as shown in **Figure 11(b)**.

Then the radiation efficiency measured result is done by using horn antenna to complete the designed antenna radiation efficiency measurement as shown in **Figure 11(b)**. In the designed antenna, the radiator and the ground plane are contributing to radiation. For UWB applications, omnidirectional radiation pattern is an important requirement. At lower frequency band of operation, the pattern resembles a conventional dipole antenna, but at higher end of the UWB spectrum,

L_{sub}	L_g	W_{sub}	W_g	W_1	W_2	W_3	S	g	W_4
50	13.5	50	24	15.3	11.7	8.5	0.9	0.6	6
L_{sep}	L_4	L_{feed}	L_3	L_1	L_2	W_f	K	P	d
7.6	2.1	45	2.8	3.6	3	6	8.5	4.5	1000

Table 2.
Dimensions of the proposed antenna (dimensions in mm) [39].

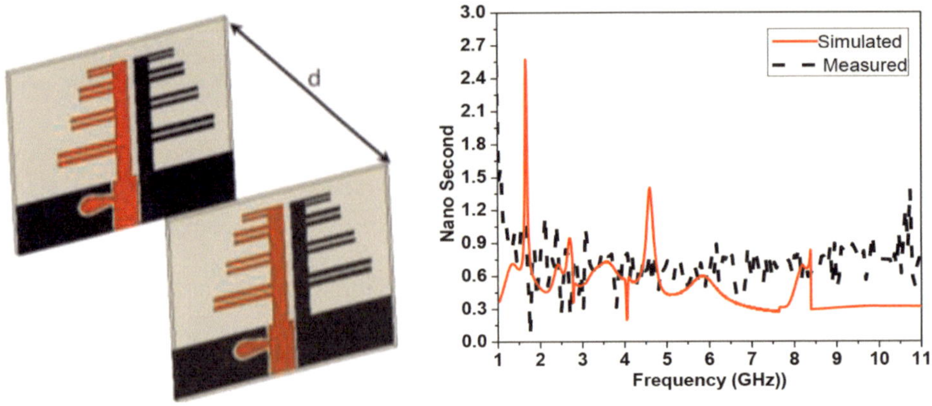

(a) (b)

Figure 10.
(a) GD simulated structures and (b) comparison between measured and simulated GD of LPDA [40].

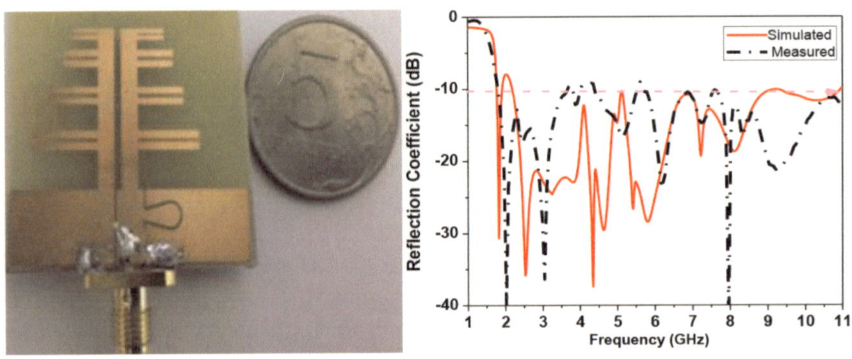

(a) (b)

Figure 11.
(a) Fabricated LPDA antenna and (b) $|S_{11}|$comparison between simulated and measured results [40].

some ripples are observed which are attributed to higher order modes. There are discrepancies observed at higher frequency band spectrum, which arises due to measurement setup.

For UWB applications, group delay is an important factor in communication systems, for example, medical applications systems, security systems, and satellite communication systems, which are used for transmitting. To avoid occurring of distortion, it is recommended that the spectrum is treated in the same manner, over the proposed bandwidth of frequencies. When GD ripples are large, they may cause unsatisfactory distortion in the signal of a transmitting radio system. So, in radio system designs, there are specifications for how much a GD may be accepted. In nonlinear systems, nonlinear distortion happens since the magnitude of frequency response is not constant, and the phase response is nonlinear. The phase distortion could be used to measure GD, the phase characteristics must have a linear slope so that the ratio is constant for all frequencies and this represents a constant GD [44]. To measure the GD between two antennas with spacing d = 1 m, the usual practice

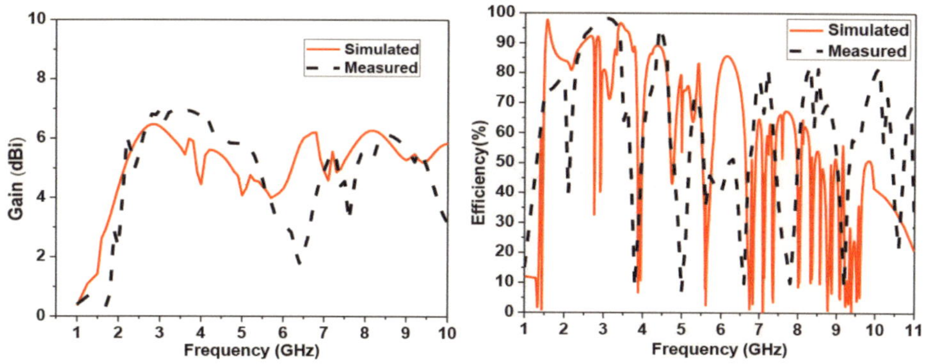

(a) (b)

Figure 12.
Comparison between simulated and measured results (a) gain and (b) radiation efficiency of LPDA [40].

is to derive Q/ω for $|S_{21}|$ phase. However, it is desirable that the same antenna is used for transition and receiving. High GD variations occur due to the steep phase shift over frequency, which may cause unsatisfactory distortions in the signal. **Figure 12** illustrates the simulated GD, and it can be noticed that the average group delay is about 1.5×10^{-9} second.

3.2 UWB monopole antenna using split ring resonator structure

This section presents designed steps to model an UWB monopole antenna. The designed antenna is composed of three different lengths of semi-circular shapes connected with circular disk and half circular modified ground plane. The designed antenna has an area equal 50×50 mm^2 on a low cost FR4 substrate [45]. The antenna demonstrates impedance bandwidth of −10 dB extended from 1.5 to 11 GHz with discontinuous bandwidth at different interior operating bands. Two pairs of SRR as metamaterial structure cells are inserted closely located from feeding transmission line of the antenna to achieve good impedance matching over the entire band of operation and improve the antenna performance. The fundamental parameters of the antenna including reflection coefficient, gain, radiation pattern, and group delay are obtained, and they meet the acceptable UWB antenna standard. Simulator HFSS ver. 14 is used as full wave electromagnetic solver then the prototypes are fabricated and measured. Results show that the antenna is very suitable for the applications in UWB as well as wireless communication systems.

For use in UWB systems, printed monopole patch antenna (PMPA) is an extremely attractive candidate because of its wide impedance bandwidth, omnidirectional azimuthal radiation pattern, low profile, and ease of integration with active devices and fabrication [46]. The design equation for lower band edge frequency has been reported in the literature. Moreover, these antennas are analyzed by using the frequency domain characteristics like return loss, gain, radiation pattern, surface current distribution, and group delay. Different narrowband services like WLAN, WiMax, GSM, UMTS, Wi-Fi, WMTS, ISM, UNII, DECT, European Hiper LAN I, II, and UWB (3.1–10.6 GHz) applications [1] could be obtained by using single UWB antenna.

A SRR is one of the electrically smallest resonant elements. It has many applications ranging from compact filters to advanced metamaterials. SRR has also a significant importance in electrically small antennas [46]. Metamaterials are good candidate for enhancement of the performances of different antennas. There are varieties of SRR structures that have been reported in the literature like square, circular, triangular, omega, and labyrinth resonator [47].

3.2.1 Antenna structure and geometry

The designed antenna structure is composed of three connected semicircular arc monopoles with circular patch fed by microstrip transmission line and modified half circular shaped ground plane as shown in **Figure 13**. The initial design is validated and optimized by simulating the proposed antenna. The proposed antenna is designed on FR4 substrate with height 1.6 mm, dielectric constant $\varepsilon_r = 4.6$, and loss tangent $\tan\delta = 0.02$. The antenna is fed by a 50-Ω transmission line (TL).

3.2.2 Design and EM models with parametric study

The main design parameter for UWB antenna is the lower frequency edge (f_L) rather than the resonance frequency (f_r) as in Eq. (7). The lower band edge frequency of the designed antenna is calculated approximately by equating their surface area with that of an equivalent cylindrical monopole antenna of the same height as given by [45]. If R_1 is the height of the planar monopole antenna in cm, which is taken the same as that of an equivalent cylindrical monopole, and r in cm is the effective radius of the equivalent cylindrical monopole antenna, which is determined by equating area of the planar and cylindrical monopole antennas, then the lower band edge frequency is given as [45]:

$$f_L = \frac{7.2}{L_f + R_1 + r} GHz \quad (7)$$

where L_f is the length of the 50 Ω feed line in cm.

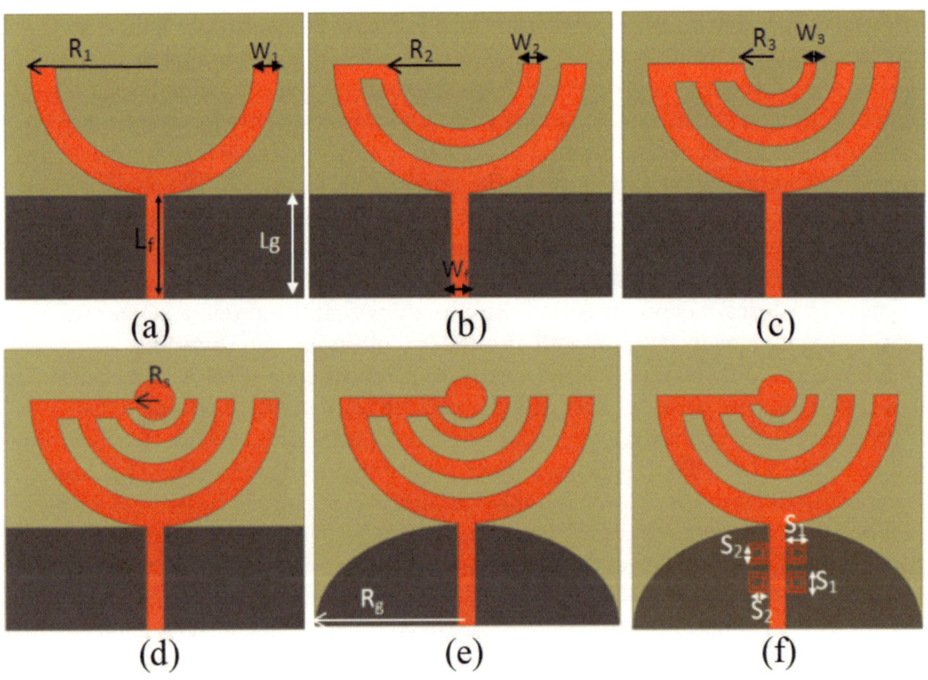

Figure 13.
Evolution of the design steps of the proposed printed monopole. (a) First step, (b) second step, (c) third step, (d) fourth step, (e) fifth step, and (f) sixth step [45].

The design started with first semi arc 180^0 with radius 25 mm and with 5 mm width modified ground plane L_g = 19 mm as shown in **Figure 13(a)**, and related $|S_{11}|$ is shown in **Figure 14**. To improve the bandwidth, second semi sector with radius 17 mm and width 3.5 mm as shown in **Figure 13(b)** with the same previous dimensions is added to add second resonant as shown in **Figure 14**. Third sector with radius 7.5 mm and width 2.5 mm is added, and keeping previous dimensions the same as shown in **Figure 13(c)**, a third resonance is achieved as shown in **Figure 14**. Another extension of the bandwidth is done by adding circular disk with radius 4 mm as shown in **Figure 13(d)**, and related reflection coefficient is shown in **Figure 14**. Modified ground plane is used to improve the bandwidth with ellipse with major radius 25 mm and minor radius 15 mm, as shown in **Figure 13(e)**, is suggested, and the related reflection is shown in **Figure 14**. The evolution of designing the proposed configuration is demonstrated in **Figure 13(f)**, and their corresponding optimized dimensions are tabulated in **Table 3**. The antenna gain and radiation efficiency are also studied as shown in **Figure 15**. The proposed antenna with SRR achieves an average gain of 7.5 dBi, and the peak realized gain around 22.5 dBi at 7.5 GHz as shown in **Figure 15(a)**. The designed antenna gain without SRR achieves an average gain around 5.5 dBi, while peak gain realized is 15 dBi at 8.5 and 10 GHz. The gain of the designed antenna is also measured, and there is good agreement between results especially at high frequency. The antenna radiation efficiency was simulated for both monopole antennas with and without SRR by using wheeler cap method [44]. The average radiation efficiency is around 70% over the operating bands for PMPA with SRR and around 65% without SRR as shown in **Figure 15(b)**.

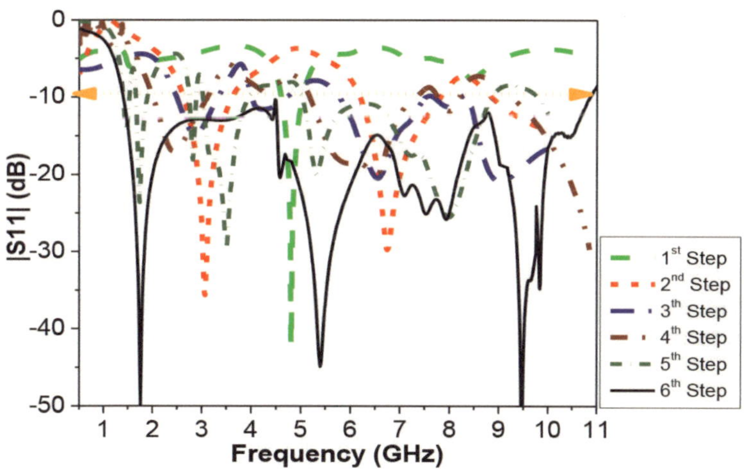

Figure 14.
Design procedures of the proposed antenna [45].

L_{sub}	W_{sub}	W_1	W_2	W_3	R_1	R_2
50	50	5	3.5	2.5	25	17
R_3	R_s	S_1	S_2	W_f	L_f	R_g
7.5	4	5	2.4	3	20	20

Table 3.
Dimensions of the proposed antenna (all dimensions in mm) [45].

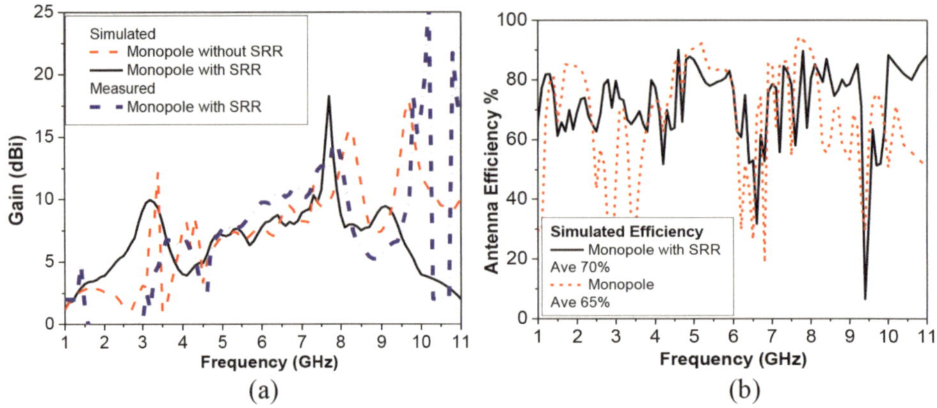

Figure 15.
(a) Antenna gain versus frequency and (b) simulated radiation efficiency with and without SRR [45].

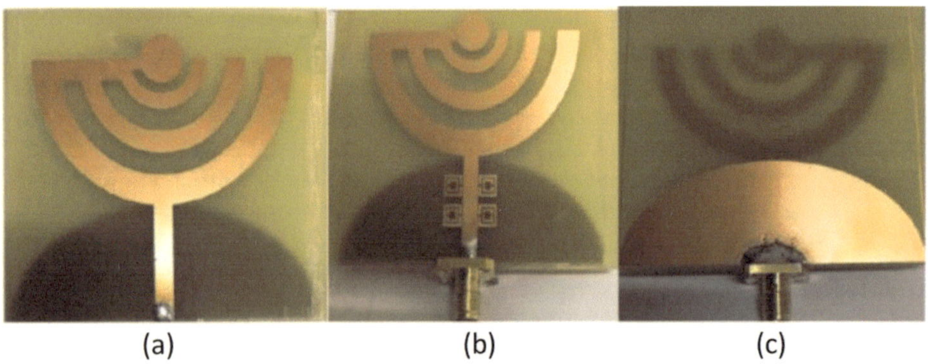

Figure 16.
Fabricated antenna (a) upper layer without SRR (b) upper layer with SRR, and (c) bottom layer [45].

3.2.3 Implementation and measured results

The prototype of the proposed antenna is shown in **Figure 16**. The performance parameters of the fabricated designed antennas are measured. The comparison results of simulated and measured input $|S_{11}|$ of the antennas are found to be in very good agreement, as shown in **Figure 17**. The −10 dB bandwidth of the designed monopole antenna with SRR extended from 1.5 to 11 GHz to cover most wireless

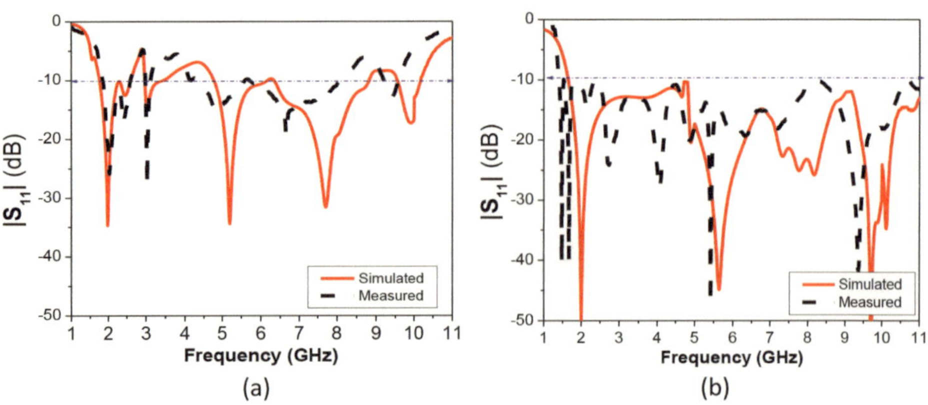

Figure 17.
Simulated and measured results of (a) monopole without SRR and (b) monopole with SRR [45].

application and FCC UWB regulation. It is fabricated by using photolithographic technique, and the measurements were carried out by using a Rohde & Schwarz ZVA67 vector network analyzer from 50 MHz to 67 GHz.

4. UWB filter

A filter is a two-port network that is used to control frequency response in a system. Filters can be classified into three main groups of active (requiring external power source), passive (no need for external power), and hybrid filters. Microwave systems are often involved with power conservation and noise control, and therefore, active filters are generally considered as last alternative. Passive filters are, however, further divided into lumped and distributed. The former consists of lumped components (including capacitors, inductors, resistors, and magnetic and electromechanical components), and the latter comprises a periodic conducting structure with various dielectric media. Inductors and capacitors form LC filters whereas resistors and capacitors form RC filters. Although resistors introduce loss to the circuit, they are generally used for broad banding (matching) purposes.

4.1 UWB band-pass filter with sharp tuned notched band rejection based on CRLH transmission-line unit cell

A compact UWB BPF with reconfigurable notch bands based on CRLH transmission line unit cell has been designed, simulated, and fabricated [48]. Two packages of software are used, namely, CST MWS and 3D EM commercial software HFSS version 13.0. The simulated and measured results are comparable. This filter has the advantage of very small size, and it also has four notched frequencies in its passband. The notched bands suppress the narrow-band services such as WLAN and WiMAX. One can control the center frequency of the notched band by varying the length L_6 of the stub. The total area of the filter is 16.4×5 mm^2. This small area makes it suitable for modern applications which need miniaturization.

4.1.1 Proposed filter design

The proposed filter is designed based on the filter described in Ref. [49] but with a new contribution which is notched controllable tunable four sharp rejection bands by adjusting the length of the coupling stub using diode switching matrix tools (instead of using PIN diodes).

Figure 18 shows the proposed microstrip UWB-BPF with tuned notched passband based on CRLH transmission-line unit cell. The optimized dimensions of the proposed filter are as shown in **Table 4**.

The dimension of the multi-mode section as shown in **Figure 18** is 4.4×1.5 mm, the length of the shunted inductive line is 3.1 mm, and the overall dimension of the proposed filter is 16.4×5.0 mm. Based on the above description, the design procedure can be as follows:

i. The notched band depends on the coupling stub (L_6) in the output section.

ii. The notch frequency of the filter can be changed by adjusting the length of the coupling stub L_6. As L_6 increases, the center notch frequency decreases as shown in **Table 5**. The length L_6 is controlled by using switching matrix equipment (mini circuit) where the character D refers to the diode.

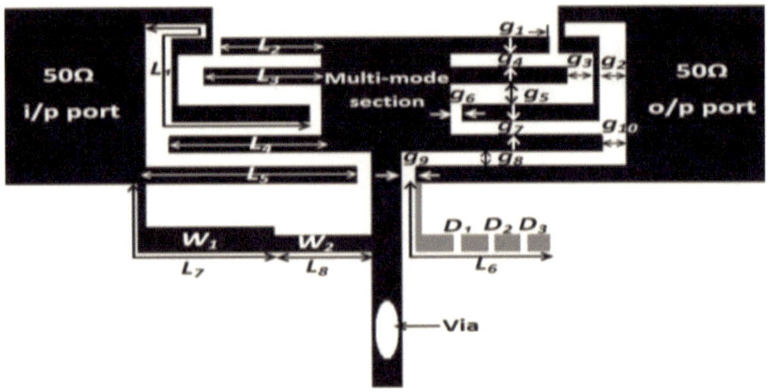

Figure 18.
The proposed filter [48].

L_1	L_2	L_3	L_4	L_5	L_7	L_8	w_1	w_2
9.3	3.4	4	5.2	7.2	5.3	3.3	0.3	0.2
g_1	g_2	g_3	g_4	g_5	g_6	g_7	g_8	g_9
0.3	0.9	0.9	0.2	0.3	0.4	0.2	0.2	0.5

Table 4.
Optimized dimensions of the proposed filter (all dimensions are in mm) [48].

Diode states	L_6 (mm)	f_{notch} (GHz)
D1, D2, D3 (off)	2	6.18
D1(on), D2, D3 (off)	3.1	5.9
D1, D2(on), D3(off)	4.2	5.7
D1, D2, D3 (on)	5.3	5.5

Table 5.
(f_{notch}) against (L_6) variation [48].

4.1.2 Fabrication and measurements

The filter was fabricated using a photolithographic technique on Rogers RT/Duroid 5880 with $\varepsilon_r = 2.2$, h = 0.787 mm, and tan $\delta = 0.0009$. The photograph of the fabricated filter is shown in **Figure 19**. The measured and simulated S_{11} and S_{21} for different stub lengths are shown in **Figure 20(a)–(d)**. The UWB bandwidth extends between 3.1 and 10.6 GHz. There are four notched frequencies for the

Figure 19.
A photo for the fabricated filter [48].

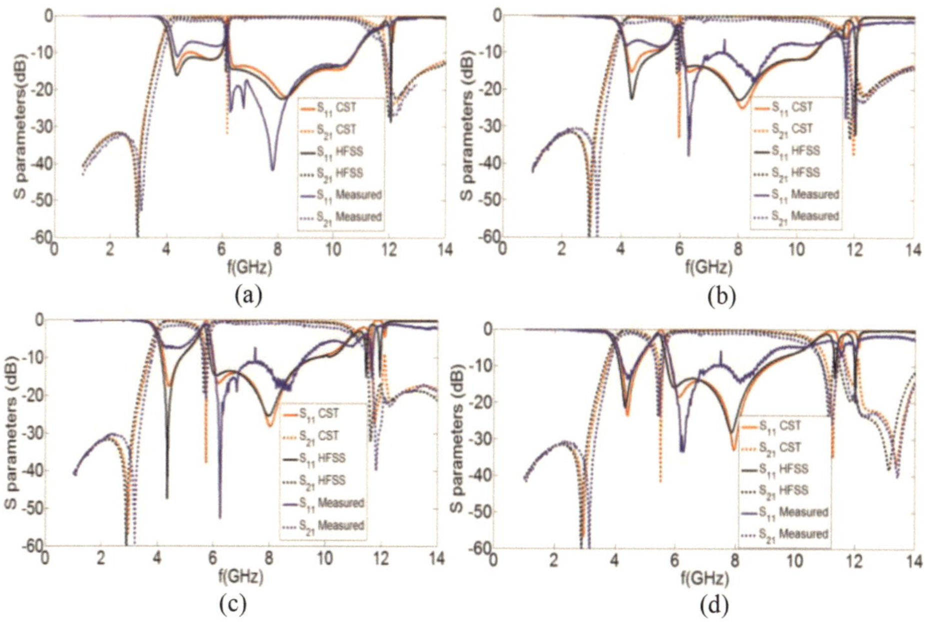

Figure 20.
The simulated and measured S_{11} and S_{21} for different L_6 lengths, (a) L_6 = 2.0 mm, (b) L_6 = 3.1 mm, (c) L_6 = 4.2 mm, and (d) L_6 = 5.3 mm [48].

different stub lengths (L_6 = 2.2, 3.1, 4.2 and 5.3 mm). The overall dimension of the filter is 16.4 × 5 mm, which is considered very compact compared to other published designs with the same characteristics. **Figure 20(a)** shows the measured and simulated S_{11} and S_{21} with L_6 = 2 mm and f_{notch} = 6.18 GHz. **Figure 20(b)** shows the measured and simulated S_{11} and S_{21} with L_6 = 3.1 mm and f_{notch} = 5.9 GHz. **Figure 20(c)** shows the measured and simulated S_{11} and S_{21} with L_6 = 4.2 mm and f_{notch} = 5.7 GHz. **Figure 20(d)** shows the measured and simulated S_{11} and S_{21} with L_6 = 5.3 mm and f_{notch} = 5.5 GHz. Good agreement was found between the measured data and simulated results.

4.2 Electronically switchable ultra-wide band/dual-band band-pass filter using defected ground structures

Ref. [50] proposed a suitable UWB to dual-band band-pass filter with defected ground structure DGS. This filter consists of four parts, namely, meandered inter digital coupled line sections, stepped impedance open stubs, coupled lines, and rectangular DGS. The filter is miniaturized and has a total area of 12.5 × 10 mm, **Figure 21**. This filter is fabricated on Duroid Teflon substrate with a dielectric constant of 2.2 and a dielectric height of 0.7874 mm. The UWB mode extends from 3.6 up to 10.6 GHz with attenuation greater than 20 dB up to 18 GHz (upper stopband). The dual passbands extended from 3.8 up to 18 GHz (upper stopband). The dual passbands extend from 3.8 up to 5 GHz and from 9.5 up to 10.8 GHz. The proposed filter suppresses WiMAX and X (military) band of satellite that extends from 7 up to 8 GHz. The filter is designed, fabricated, and measured. The mode of the filter is changed by using suitable matrix equipment [50].

DGS at input and output ports of the proposed filter produces two resonances at 7.5 and 9.6 GHz and improve the performance of proposed filter while an overall size reduction of 20% is obtained. The meander lines and stepped impedance open stub are also used to reduce the overall size. By adjusting the connection between

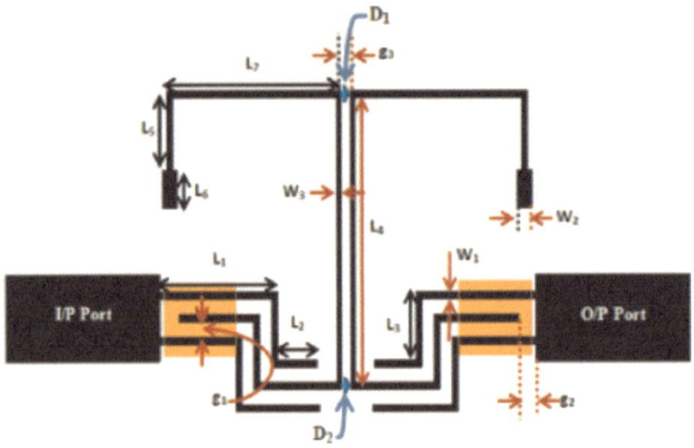

Figure 21.
The structure of the proposed filter [50].

the coupled lines in the center of the design, the center frequency and 3 dB frequency band can be easily adjusted. The proposed filter achieves UWB performance with good selectivity and low insertion loss in the passband from 3.6 to 10.5 GHz and good stopband from 10.6 to 18 GHz. Moreover, it achieves dual bands with good stopband from 5 to 9.5 GHz and from 10.8 to 18 GHz by using open circuit stub to suppress unwanted interference signals in the band of WLAN, WIMAX, and X (Military) band of satellite. All dimensions of the proposed filter are as follows: L_1 = 3.75 mm, L_2 = 1.95 mm, L_3 = 1.8 mm, L_4 = 7.5 mm, L_5 = 2.1 mm, L_6 = 1 mm, L_7 = 5.65 mm, W_1 = 0.2 mm, W_2 = 0.5 mm, W_3 = 0.15 mm, g_1 = 0.2 mm, g_2 = 0.2 mm, and g_3 = 0.3 mm. The simulated S_{11} and S_{21} are shown in **Figure 24**.

4.2.1 The equivalent lumped circuit model analysis of the proposed design

Figure 22 shows the equivalent lumped circuit model of the proposed UWB BPF that is shown in **Figure 21**. The equivalent lumped circuit model results are obtained using circuit model tool of the Advanced Design System (ADS) 2017. The lumped element values are manually optimized by changing each element value, so that it can have good agreement with the simulated results obtained from the full wave simulator.

The whole equivalent circuit of the proposed filter can be divided into the following subsections: DGS part at input and output ports, interdigital coupled lines and stepped impedance open stub. As shown in the lumped element model (**Figure 22**), R_{d1}, C_{d1}, L_{d1}, R_{d2}, C_{d2}, and L_{d2} represent the equivalent resistance,

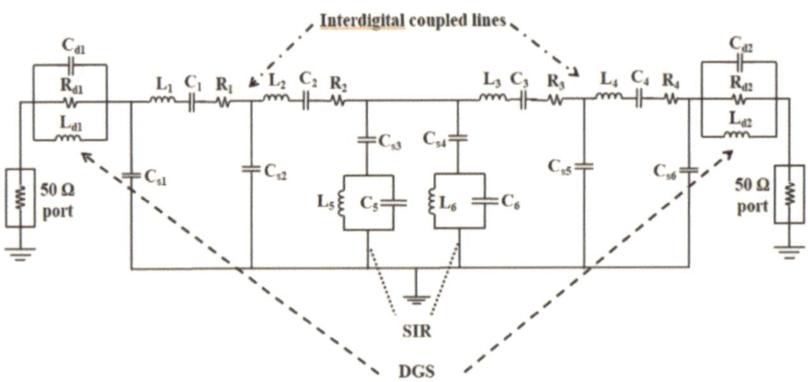

Figure 22.
Equivalent lumped circuit model of the proposed UWB BPF shown in ***Figure 21*** *[50].*

inductance, and capacitance of the defected ground structure (DGS) [51]. L_5, C_5, L_6, and C_6 represent the equivalent inductance and capacitance of the stepped impedance resonator (SIR). Interdigital coupled arm is represented by the series capacitance with parasitic inductance and resistance, and shunt capacitances [52] as shown in **Figure 22**. The S-parameters versus frequency response of EM simulation and circuit model are compared. There is a very good agreement between the simulated and equivalent lumped circuit model results.

4.2.2 Fabrication and measurements

Photolithographic technique was used to fabricate this filter on Teflon substrate (Duroid RT 5880) with physical properties of ε_r =2.2 and $\tan\partial$ = 0.0009, while the dielectric thickness is 0.7874 mm. **Figure 23** shows a photograph for the fabricated filter for both sides (the front and back sides). The soldered wires shown in **Figure 23** are used to connect the filter with diode switch matrix tool. The filters are measured using the vector network analyzer (N9928A FieldFox Handheld Microwave Vector Network Analyzer, 26.5 GHz) [50].

Figure 24(a) shows the measured and simulated results of the proposed filter at ON state with frequency range from 1 to 20 GHz. It should be noted that the frequency range is extended up to 20 GHz in order to show that the out of band rejection is good, and the measured 3 dB passband of the proposed filter is between 3.6 and 10.6 GHz. **Figure 24(b)** shows the measured and simulated results of the proposed filter at OFF state, and the dual bands with 3 dB passbands extend from 3.8 to 5 GHz and from 9 to 10.8 GHz [50].

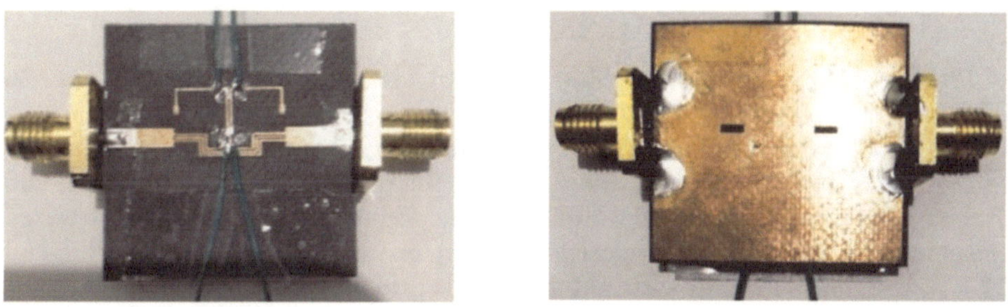

Figure 23.
A photo for the fabricated filter [50].

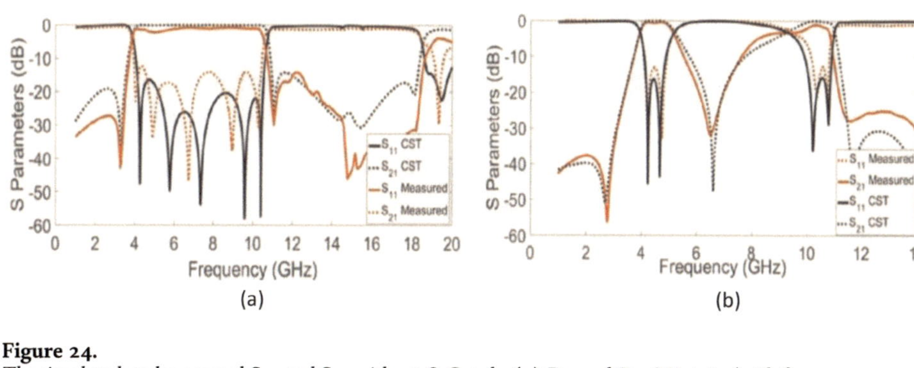

Figure 24.
The simulated and measured S_{11} and S_{21} without O.C stub. (a) D1 and D2 ON state (with frequency range from 1 to 20 GHz) and (b) D1 and D2 OFF state [50].

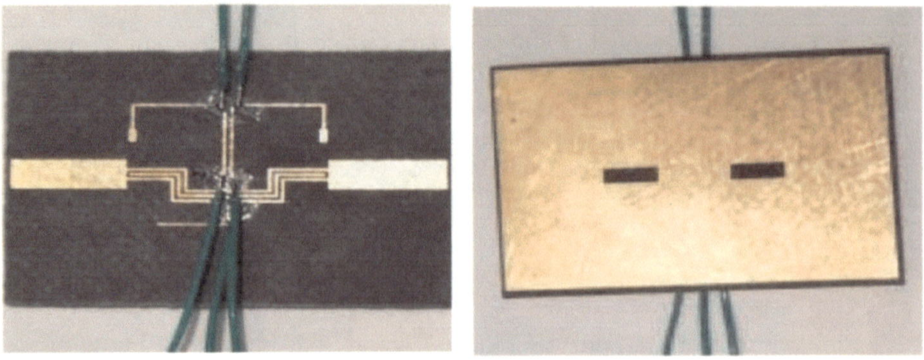

Figure 25.
*A photo for the fabricated filter of **Figure 10**.*

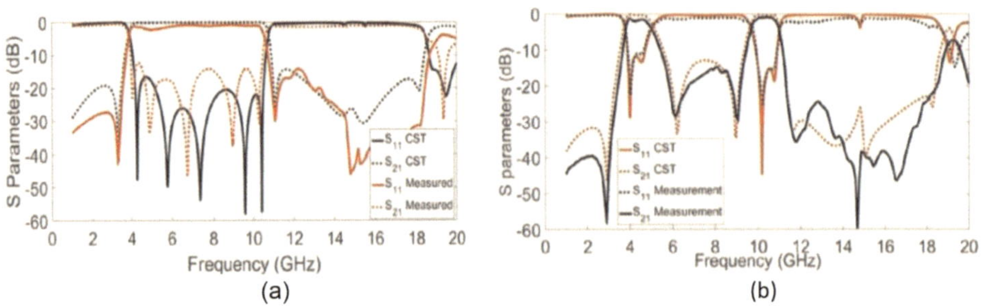

Figure 26.
The simulated and measured S_{11} and S_{21} with O.C stub. (a) D1, D2 ON, and D3 OFF, (b) D1, D2 OFF and D3 ON [50].

Photos for the fabricated filter with open stub are shown in **Figure 25**. **Figure 26 (a)** shows the measured and simulated results of the proposed filter with open stub at D1, D2 ON state, and D3 OFF with frequency range from 1 to 20 GHz. It should be noted that the out of band rejection is good, and the measured 3 dB passband of the proposed filter is between 3.6 and 10.6 GHz. **Figure 24(b)** shows the measured and simulated results of the proposed filter with open stub at D1, D2 OFF state and D3 ON, and the dual bands with 3 dB passbands extend from 3.8 to 5 GHz and from 9.5 to 10.8 GHz [50].

5. UWB filtenna

In general, the filtenna consists of a filter and antenna that are combined in one structure. The proposed filtenna operates at three bands of frequency (2.4, 5.5, and 28 GHz) to cover the 4G/5G communication system. It consists of three parts, namely, Franklin strip monopole antenna to cover 4G, WLAN, and WiMAX and a rectangular patch antenna to cover 5G band. The third part consists of a modified CMRC low-pass filter that exists between the two antenna parts to isolate the Franklin antenna from the rectangular patch antenna at 5G band. It also allows feeding the Franklin antenna at low frequency bands. The total size of the filtenna is $45 \times 40 \times 0.508$ mm^3 and fabricated on Teflon dielectric substrate (Roger 5880). The proposed filtenna has wide impedance bandwidth (15.8, 23.5, and 11.3%) and high gain (1.95, 3.76, 7.35 dBi) [53]. The proposed multiband filtenna is shown in **Figure 27**.

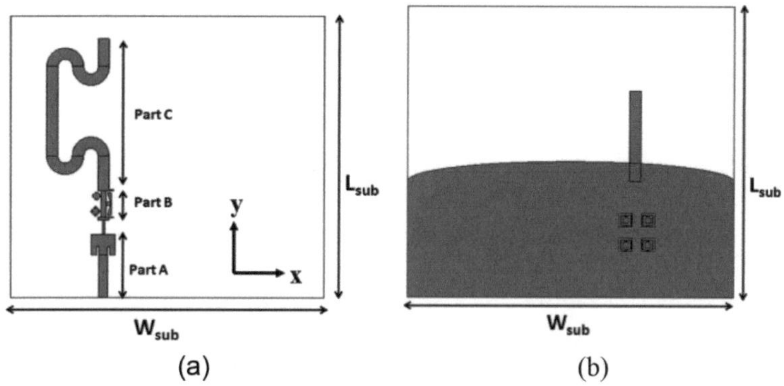

Figure 27.
The proposed multiband filtenna (a) front view and (b) back view [53].

5.1 Modified CMRC LPF using novel fractal patches

A modified compact microstrip resonance cell (CMRC) low-pass filter (LPF) using novel fractal patches was proposed in [54], see **Figure 28**. The fractal patches produce additional transmission zeros to the stop-band, while the open-ended stubs cause an extension in the stopband achieving a compact ultrawide and deep stopband filter with good selectivity and low insertion loss in the passband. The results show −10 dB bandwidth from 3.3 to 67 GHz with 181.5% relative stopband bandwidth. The 3-dB cutoff frequency is 2.85 GHz and less than 1.5 dB insertion loss in the passband and 0.55 GHz transaction from −3 to −20 dB and − 20 dB suppression from 3.5 to 67 GHz, so that the filter can be expected to suppress the unwanted harmonics and prevent inter-modulation with the new systems with high frequency operating bands. The filter has been designed on a Rogers 5880 substrate with a relative dielectric constant of 2.2, substrate thickness of 0.508 mm, and 0.0009 loss tangent. **Figure 28** shows the proposed filter design, and it consists of two traditional triangle taper resonance cells in one side of the transverse connecting narrow width transmission line which has almost the same performance of the complete CMRC structure, while two different sizes circular fractal patches are present on the other half. Each fractal consists of main circular patch and additional small circular patches at edges. The two fractals act as a dual behavior resonator to have additional transmission zeros in the stopband. Each fractal is resonating at certain frequency in addition with enhancing the low suppression bands of the entire stop-band. Also, four open ended stubs are used to extend the

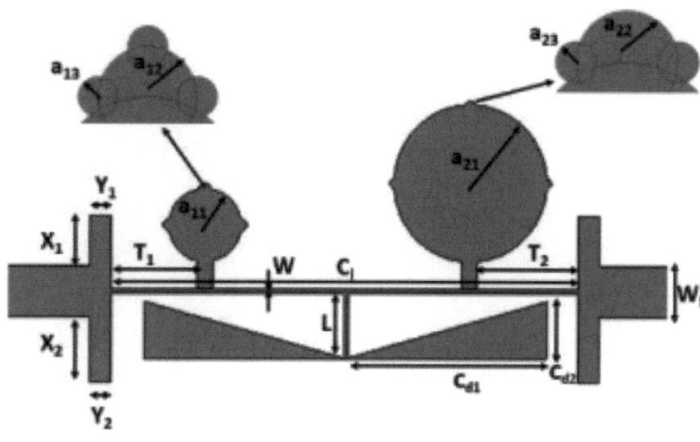

Figure 28.
The design of proposed low-pass filter [53].

stopband by adding new transmission zeros without increasing the circuit size. The main dimensions are given in **Table 6**, all dimensions in millimeter.

5.2 Franklin monopole antenna

Part C of the multiband rectenna system is a Franklin strip monopole dual-band antenna to be used to cover Bluetooth at 2.4 GHz, 4G, LTE bands at 2.3, 2.5 and 2.6 GHz, WiMAX at 2.5 and 5.5GHz, WLAN at 2.4, 5.2 GHz [53]. The geometry of the antenna is shown in **Figure 29**. The antenna has a rectangular stub on a curved partial ground. The length of the bending strip is about one-half of the guide wavelength at its first resonance frequency. The meander radius and the length of the Franklin strip are mainly determining the two resonance frequencies of the antenna, while the rectangular stub with a length of a quarter wavelength and curved ground have been used to increase the bandwidth of the upper band (5.5 GHz). The L-C equivalent circuit of the Franklin monopole antenna is shown in **Figure 30**. The dimensions of the antenna are shown in **Table 7**, while the equivalent circuit parameters are shown in **Table 8**.

5.3 Rectangular patch antenna

The first part of the rectenna (part A) consists of a rectangular patch antenna with inset feed for matching and four CSRRs (complementary split ring resonator) in the other side (ground plane). This antenna covers the 5G range of frequency (28 GHz). This shape is chosen due to its simplicity and can be placed in the Franklin feeding line. **Figure 31** shows the geometry of the rectangular patch with inset feeding. The final dimensions of the antenna after using optimization techniques of the CST simulator are introduced in **Table 9**. The L-C equivalent circuit of this antenna with CSRRs is shown in **Figure 32**. The rectangular patch is

Parameter	C_l	L	W	C_{d1}	C_{d2}	a_{11}	a_{12}	a_{13}	a_{21}
Value (mm)	17.6	2.4	0.15	7.6	2.1	1.35	0.2	0.05	2.9
Parameter	a_{22}	a_{23}	X_1	X_2	Y_1	Y_2	T_1	T_2	W_{50}
Value (mm)	0.15	0.05	2.4	1.8	0.8	0.8	3.2	3.8	1.6

Table 6.
Circuit dimensional parameters [53].

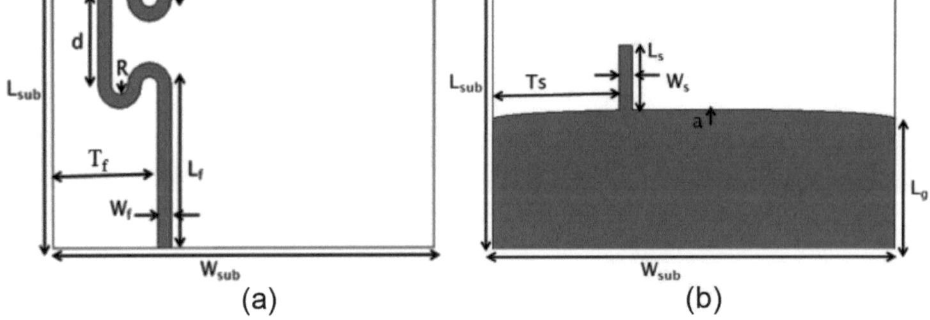

Figure 29.
The design of the proposed Franklin strip monopole antenna (a) front and (b) back [53].

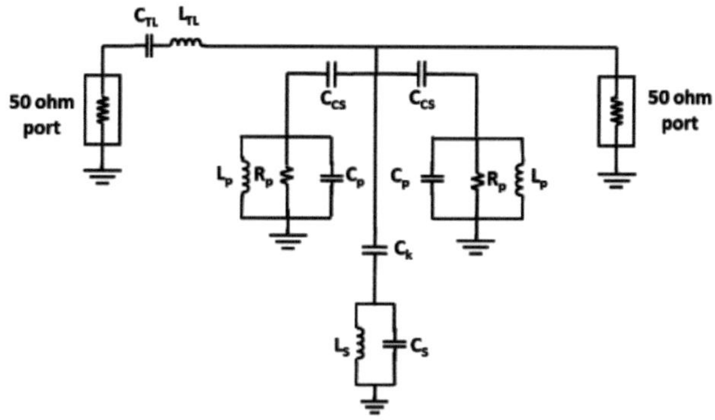

Figure 30.
The equivalent LC circuit of the proposed 4G Franklin monopole antenna,

L_{sub}	W_{sub}	L_f	L_g	L	d
35	45	18.5	14.9	5.1	9.9
R	**a**	**W_f**	**W_s**	**T_f**	**T_s**
1	1.1	1.6	1.5	13.4	14.9

Table 7.
Dimensions of the proposed Franklin antenna (all dimensions in mm) [53].

CTL	LTL	Cp	Lp	Rp	CS	LS	Ck	Ccs
2	1.7	2606	0.0007	0.05	0.1	0.05	18.4	42.4
C in Pico Farad, L in Nano Henry and R in ohm.								

Table 8.
Franklin monopole antenna equivalent circuit parameters [53].

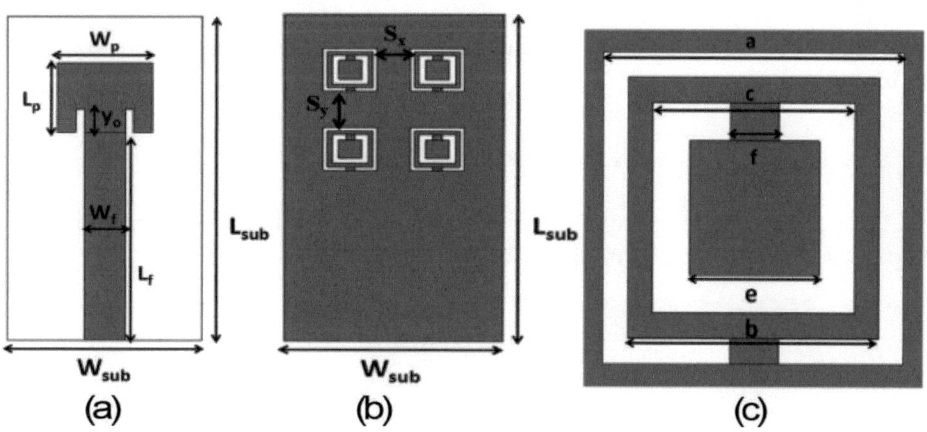

Figure 31.
The design of 5G rectangular patch antenna. (a) Front view, (b) back view, and (c) the dimensions of the CSRR [53].

represented as two radiating slots each one represented by a parallel R L C circuit, while each one of the four CSRRs represented by parallel LC and the electric coupling between the CSRRs on the ground and radiating patch on the top side is represented by capacitor C_k and the magnetic coupling represented by the mutual

L_{sub}	W_{sub}	L_f	W_f	y_o	L_p	W_p
11	7.5	6	1.6	1	2.95	3.45
S_x	S_y	**a**	**b**	**c**	**e**	**f**
1.15	1.55	1.85	1.55	1.25	0.8	0.3

Table 9.
5G antenna dimensional parameters (all dimensions in mm).

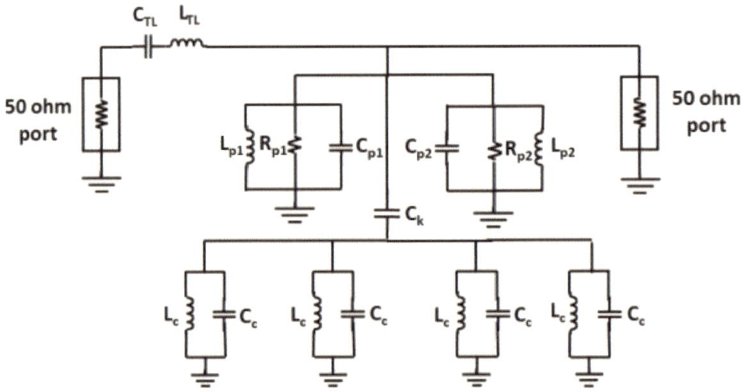

Figure 32.
(a) The equivalent LC circuit of the proposed 5G rectangular patch antenna [53].

C_{TL}	L_{TL}	C_{p1}	L_{p1}	R_{p1}	C_{p2}	L_{p2}	R_{p2}	C_c	L_c
17.5	0.16	1.06	0.26	93.6	10^{-12}	0.02	54.9	0.79	1.2
C in Pico farad, L in Nano Henry and R in ohms.									

Table 10.
5G antenna equivalent circuit parameters.

inductance offered by the ADS software. All lumped element values are listed in **Table 10** [53].

5.4 Simulation and measurement results of filtenna

The equivalent circuit model simulation results of the filtenna system shown in **Figure 27** is determined by merging the three parts component's equivalent circuit,

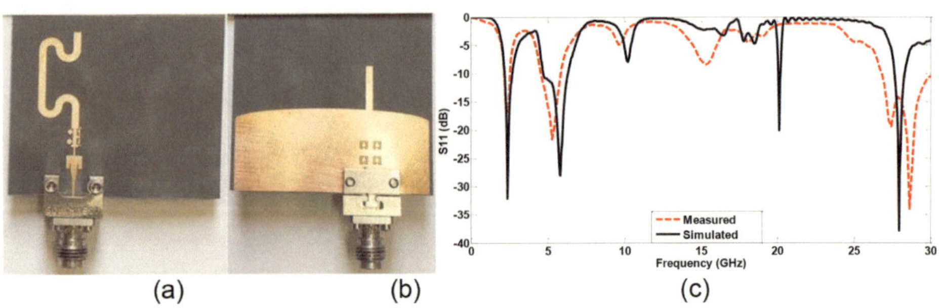

Figure 33.
Photograph of fabricated antenna. (a) Front view, and (b) back view and (c) the measured and simulated reflection coefficient [53].

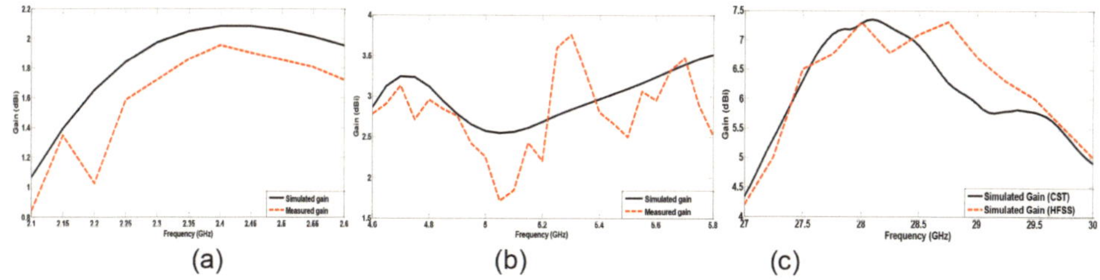

Figure 34.
Simulated and measured gain at: (a) first band, (b) second band, and (c) simulated gain using CST and HFSS simulators for the third band [53].

namely, 5G rectangular patch antenna, modified CMRC low-pass filter, and 4G Franklin monopole antenna extract using ADS software [53]. The filtenna was fabricated using the photolithographic technique.

Figure 32 shows a photo of the fabricated rectenna. The simulated and measured reflection coefficient is shown in **Figure 33**. The measured results show that the filtenna has −10 dB impedance of the first band from 2.16 to 2.53GHz, the second band is from 4.58 to 5.8GHz, and the third band is from 26.8 GHz to 30 GHz. The simulated and measured gains are shown in **Figure 34**. The first and second bands have peak measured gain level of 1.95 and 3.76 dBi, respectively. The third band achieves 7.35 dBi peak simulated gain level [53].

Author details

Dalia N. Elsheakh* and Esmat A. Abdallah
Microstrip Department, Electronics Research Institute, Giza, Egypt

*Address all correspondence to: daliaelsheakh@gmail.com

References

[1] Federal Communication Commission. First Report and Order-Revision of Part 15 of the Commission's Rules Regarding Ultra-Wideband Transmission System. FCC 02 48; 2002

[2] Zhang Q. Ultra-Wideband Impulse Radio Communication System Design and Prototyping. USA: Tennessee Technological University;

[3] Qiu RC, Liu HP, Shen X. Ultra-wideband for multiple access. IEEE Communications Magazine. 2005;**43**: 80-87

[4] Qiu RC. A study of the ultra-wideband wireless propagation channel and optimum UWB receiver design, Part I. IEEE Journal on Selected Areas in Communications. 2002;**20**(9):1628-1637

[5] Chong C-C, Watanabe F, Inamura H. Potential of UWB technology for the next generation wireless communications. In: IEEE Ninth Inter-National Symposium on Spread Spectrum Techniques and Applications; Manaus, Amazon, Brazil. 2006

[6] Fakharzadeh M, Mohajer M. An integrated wide-band circularly polarized antenna for millimeter-wave applications. IEEE Transactions on Antennas and Propagation. 2014;**62**: 925-929

[7] Pinel S, Kim K, Yang K, Laskar J. 60 GHz linearly and circularly polarized antenna arrays on liquid crystal polymer substrate. In: Proceedings of 36th European Microwave Conference. 2006. pp. 858-861

[8] Yeung SH, Man KF, Chan WS. A bandwidth improved circular polarized slot antenna using a slot composed of multiple circular sectors. IEEE Transactions on Antennas and Propagation. 2011;**59**: 3065-3070

[9] James JR, Hall PS, editors. Handbook of Microstrip Antennas. London: Peter Peregrinus; 1989

[10] Hall PS, Dahele JS. Dual and circularly polarized microstrip antennas. In: Lee KF, Chen W, editors. Advances in Microstrip and Printed Antennas. New York: Wiley; 1997

[11] Hsu CL, Hsu FC, Kuo JT. Microstrip bandpass filters for ultra-wideband (UWB) wireless communications. In: IEEE MTT-S International Microwave Symposium Digest. 2005. pp. 679-682

[12] Hong JS, Shaman H. An optimum ultra-wide-band microstrip filter. Microwave and Optical Technology Letters. 2005;**47**:230-233

[13] Li X, Ji X. Novel compact UWB bandpass filters design with cross-coupling between short-circuited stubs. IEEE Microwave and Wireless Components Letters. 2014;**24**(1):23-25

[14] Zhu L, Sun S, Menzel W. Ultra-wideband (UWB) bandpass filters using multiple-mode resonator. IEEE Microwave and Wireless Components Letters. 2005;**15**(11):796-798

[15] Xu J, Wu W, Kang W, Miao C. Compact UWB bandpass filter with a notched band using radial stub loaded resonator. IEEE Microwave and Wireless Components Letters. 2012;**22**(7):351-353

[16] Shaman H, Hong JS. Ultra-wideband (UWB) bandpass filter with embedded band notch structure. IEEE Microwave and Wireless Components Letters. 2007;**17**(3):193-195

[17] Kim CH, Chang K. Ultra-wideband (UWB) ring resonator bandpass filter with a notched band. IEEE Microwave and Wireless Components Letters. 2011; **21**(4):206-208

[18] Wong SW, Zhu L. Implementation of compact UWB bandpass filter with a notch-band. IEEE Microwave and Wireless Components Letters. 2008; **18**(1):10-12

[19] Yang G-M, Jin R, Vittoria C, Harris VG, Sun NX. Small ultra-wideband (UWB) bandpass filter with notched band. IEEE Microwave and Wireless Components Letters. 2008; **18**(3):176-178

[20] Hao Z-C, Hong J-S, Parry JP, Hand DP. Ultra-wideband bandpass filter with multiple notch bands using nonuniform periodical slotted ground structure. IEEE Transactions on Microwave Theory and Techniques. 2009;**57**(12):3080-3088

[21] García RG, Guyette AC. Reconfigurable multi-band microwave filters. IEEE Transactions on Microwave Theory and Techniques. 2015;**63**(4): 1294-1307

[22] Wang H, Tam KW, Ho SK, Kang W, Wu W. Design of ultra-wideband bandpass filters with fixed and reconfigurable notch bands using terminated cross-shaped resonators. IEEE Transactions on Microwave Theory and Techniques. 2014;**62**(2):252-265

[23] Horestani AK, Shaterian Z, Naqui J, Martín F, Fumeaux C. Reconfigurable and tunable S-shaped split-ring resonators and application in band-notched UWB antennas. IEEE Transactions on Antennas and Propagation. 2016;**64**(9):3766-3776

[24] Liang JG, Zhang X, Sun L. Compact UWB bandpass filter with triple notched bands using quadruple-mode resonator. In: IEEE International Conference on Microwave and Millimeter Wave Technology (ICMMT). Vol. 1. Beijing, China; 2016. pp. 354-356

[25] Zheng X, Liu W, Zhang X, Jiang T. Design of dual band-notch UWB bandpass filter based on T-shaped resonator. In: Progress in Electromagnetic Research Symposium; 8-11 August 2016; Shanghai, China. 2016. pp. 4482-4486

[26] Ahmed KU, Virdee BS. Ultra-wide band bandpass filter based on composite right/left handed transmission-line unit cell. IEEE Transactions on Microwave Theory and Techniques. February 2013; **61**(2):782-788

[27] Elsheakh DN. Linear/circular polarization slot antennas for millimeter wave application. Microwave and Optical Technology Letters. 2017;**59**(4): 976-983

[28] Madhuri S, Tiwari VN. Review of circular polarization techniques for design of microstrip patch antenna. Conference: International Conference on Recent Cognizance in Wireless Communication & Image Processing, Jaipur, 2014

[29] Pozar DM, Schaubert DH. Microstrip Antennas: The Analysis and Design of Microstrip Antennas and Arrays. New York: IEEE Press; 1995

[30] Nesic D, Nesic A, Brankovic V. Circular polarized printed antenna with broadband axial ratio. In: Presented at the IEEE Antennas Propagation Society International Symposium. 2003. pp. 912-915

[31] Li RL, Dejean G, Laskar J, Tentzeris MM. Investigation of circularly polarized loop antennas with a parasitic element for bandwidth enhancement. IEEE Transactions on Antennas and Propagation. 2005;**53**: 3930-3939

[32] Weily AR, Guo YJ. Circularly polarized ellipse-loaded circular slot array for millimeter-wave WPAN applications. IEEE Transactions on Antennas and Propagation. 2009;**57**: 2862-2870

[33] Wong K-L. Compact and Broadband Microstrip Antennas. 1st ed. New York: Wiley-Interscience; 2002

[34] Sun H, Guo YX, Wang Z. 60-GHz circularly polarized U-slot patch antenna array on LTCC. IEEE Transactions on Antennas and Propagation. 2013;**61**:430-435

[35] Nakano H, Eto J, Okabe Y, Yamauchi J. Tilted and axial beam formation by a single-arm rectangular spiral antenna with compact dielectric substrate and conducting plane. IEEE Transactions on Antennas and Propagation. 2002;**50**:17-23

[36] Bayat A, Mirzakhani R. A parametric study and design of the balanced antipodal Vivaldi antenna (BAVA). In: PIERS Proceedings; 19-23 August 2012; Moscow, Russia. 2012. pp. 778-782

[37] Agahi MH, Abiri HL, Mohajeri F. Investigation of a new idea for antipodal Vivaldi antenna design. International Journal of Computer and Electrical Engineering. 2011;**3**(2):1793-8163

[38] Zhang F, Fang G-Y, Ji YC, Ju H-J, Shao J-J. A novel compact double exponentially tapered slot antenna (DETSA) for GPR applications. IEEE Antennas and Wireless Propagation Letters. 2011;**10**:195-198

[39] Elsheakh DN, Abdallah EA. Compact printed log-periodic dipole antenna for terrestrial digital video broadcast (DVB-T) application. Microwave and Optical Technology Letters. 2014;**56**:1002-1007

[40] Lin S, Luan S, Wang YD, Luo X, Han X, Zhang XQ, et al. A printed log periodic tree dipole antenna (PLPTDA). Progress in Electromagnetics Research M. 2011;**21**:19-32

[41] Wu J, Zhao Z, Nie Z, Liu QH. A printed UWB vivaldi antenna using stepped connection structure between slot line and tapered patches. IEEE Antennas and Wireless Propagation Letters. 2014;**13**:698-701

[42] Yu C, Hong W, Chiu L, Zhai G, Yu C, Qin W, et al. Ultrawideband printed log-periodic dipole antenna with multiple notched bands. IEEE Transactions on Antennas and Propagation;**59**:725-732

[43] Casula G, Maxia P, Mazzarella G, Montisci G. Design of a printed log periodic dipole array for ultra-wideband applications. Progress in Electromagnetics Research C. 2013;**38**:15-26

[44] Pozar DM, Kaufman B. Comparison of three methods for the measurement of printed antenna efficiency. IEEE Transactions on Antennas and Propagation. 1988;**36**:136-139

[45] Elsheakh DM, Abdallah EA. Ultra-wide-bandwidth (UWB) microstrip monopole antenna using split ring resonator (SRR) structure. International Journal of Microwave and Wireless Technologies. 2018;**1**:1-10

[46] Erentok A, Ziolkowski RW. Metamaterial-inspired efficient electrically small antennas. IEEE Transactions on Antennas and Propagation. 2008;**56**:691-707

[47] Erentok A, Ziolkowski RW. Two-dimensional efficient metamaterial-inspired electrically small antenna. Microwave and Optical Technology Letters. 2007;**49**:1669-1673

[48] Ouf EG, Mohra AS, Abdallah EA, Elhennawy H. Ultra-wideband bandpass filter with sharp tuned notched band rejection based on CRLH transmission-line unit cell. Progress in Electromagnetics Research Letters. 2017;**69**:9-14

[49] Ahmed KU, Virdee BS. Ultrawide band bandpass filter based on composite

right/left handed transmission line unit cell. IEEE Transactions on Microwave Theory and Techniques. 2013;**61**(2): 782-788

[50] Ouf EG, Mohra AS, Abdallah EA, Elhennawy H. Electronically switachable ultra-wide band/dual-band bandpass filter using defected ground structures. Progress in Electromagnetics Research C. 2019;**91**:83-96

[51] Sarkar D, Moyra T, Murmu L. An ultra-wideband (UWB) bandpass filter with complementary split ring resonator for coupling improvement. International Journal of Electronics and Communications. 2017;**71**:89-95

[52] Weng LH, Guo Y-C, Shi X-W, Chen X-Q. An overview on defected ground structure. Progress in Electromagnetics Research B. 2008;**7**: 173-189

[53] Yassin ME, Mohamed HA, Abdallah EA, Elhennawy H. Single fed 4G/5G multiband 2.4/5.5/28GHz antenna. IET Microwaves, Antennas and Propagation. March 2018;**13**: 286-290

[54] Yassin ME, Mohamed HA, Abdallah EA, Elhennawy H. Modified CMRC LPF using novel fractal patches. Progress in Electromagnetics Research Letters. 2018;**79**:25-31